ORIGIN 2.0

FROM THE BIG BANG TO THE FUTURE OF HUMANITY

W. MARC POSTLEWAITE

Wonder Books
Press

For Marion, with gratitude for her patience, steady support, and unshakable faith in me.

"Origin 2.0: the story of minds made of matter
—fragile, curious, and responsible."

CONTENTS

PREFACE

I wrote this book initially as an exercise: to study the origins of the universe and to write down my research so I could better understand it. As I progressed, the questions grew larger and began to feed on one another — a feedback loop of curiosity that pulled the story outward from atoms to life, from life to minds, and from minds to the institutions and technologies that now shape our shared future.

Before going further, a clarification of scope. This is a book about the *scientific* story of creation — the origins of matter, life, complexity, societies, and the challenges we now face. It is not a religious account of creation, nor does it attempt to evaluate or compare the creation narratives found in various religious beliefs. Those narratives have their own deep and rich histories and thousands of books devoted to them. This book does not enter that discussion; it simply follows the scientific arc alone.

This book is also not meant to offer an exhaustive, technical exploration of every physical, chemical, or biological stage of evolution. In truth, each chapter could expand into several volumes on its own. Instead, this is a fast journey across 13.8 billion years, distilled into 19 chapters. I think of it like a speedboat racing from

Point A to Point B. It skims the surface, hitting only the high spots of the waves, never diving too deep — because doing so would slow the journey. In the same way, we will keep our momentum. We will not plunge into every scientific detail of each evolutionary milestone; we'll glide across the surface, touching the high points as we travel through time.

Along the way, I occasionally include personal reflections or philosophical observations. These are neither revelations nor claims of profound wisdom — simply thoughts that emerged while trying to connect many branches of science into one accessible story. This book makes no claim to authority beyond the science itself; its goal is simply to make the scientific story understandable to curious non-scientists.

Origin 2.0 is not a textbook, a technical monograph, or a manifesto. It is an attempt to trace a long arc — cosmology to chemistry, chemistry to biology, biology to culture, and culture to technology — in language accessible to an intelligent reader who is not a specialist in any one field. My aim is practical and modest: to provide readers with a basic understanding of how the universe evolved, equipping them with the conceptual grounding to judge the serious issues confronting our world now and in the future— and to engage in intellectual, meaningful conversations about them.

A few points of method and limitation up front. Where scientific consensus is strong, I summarize it; where the literature is unsettled, I present plausible ranges and the key assumptions that move them. The material on climate and on artificial intelligence is necessarily selective: both fields evolve rapidly, and any snapshot will date. That is why this book emphasizes durable principles — architectures of governance, institutional design, and moral reasoning — rather than betting on specific dates or singular technical fixes.

The book is organized so you can read it as a single narrative or dip into the parts that interest you most. The early chapters tell a condensed origin story: how matter and energy conspired to create environments hospitable to complexity, how self-replication and imperfect copying allowed life to vary and adapt, and how shared narratives turned scattered minds into cooperating societies. The later sections turn to the tests those societies now face: an atmosphere being altered in ways that risk irreversible changes, and technologies that amplify human power while inheriting our strengths and our faults.

Two themes keep circling back. The first is simple: mortality matters. Our limited time is not simply a boundary—it is the well-spring of prudence, sacrifice, and responsibility. Recognizing that gives moral weight to stewardship — the choice to protect futures we may not personally see. Second, constraints matter. Powerful systems must be designed with limits: technical roots of trust, institutional checks, transparency, and democratic participation. Without those constraints, tools that could help solve collective problems may instead become vectors of harm.

This book is written as an invitation. The problems it treats are collective problems; their solutions will be collective as well. If these pages sharpen your questions, nudge you to discuss them with neighbors, or inspire small practical steps — from civic engagement to supporting better governance of technology and energy — then the book will have done useful work.

A final note about tone: I have tried to be neither alarmist nor complacent. The stakes are high, but clarity about the problems and the levers we possess is a form of power. Read critically, argue vigorously, and take part in the slow democratic work of deciding what sort of world we want to pass on.

— **Marc Postlewaite**

1 FROM NOTHING TO SOMETHING

The most incomprehensible thing
about the universe is that it is comprehensible
—Albert Einstein
(Theoretical Physicist)

The Incomprehensible Beginning

We begin this story with something so small and so strange that our normal ideas of size don't even make sense. It didn't have width, or weight, or mass the way we think of them — and yet, hidden within it were all the ingredients needed to make every star, planet, and living thing. This was the universe 13.8 billion years ago, in a state so dense and so hot that the laws of physics as we know them could not describe it. Scientists call this mysterious beginning a **singularity** — not because it was a tiny dot in space, but because it was a moment when space, time, and the rules of nature had not yet taken shape.

From that beginning, the universe expanded with astonishing speed.

What was once smaller than anything we can imagine has now

grown into something larger than anything we can imagine: the vast cosmos we see today.

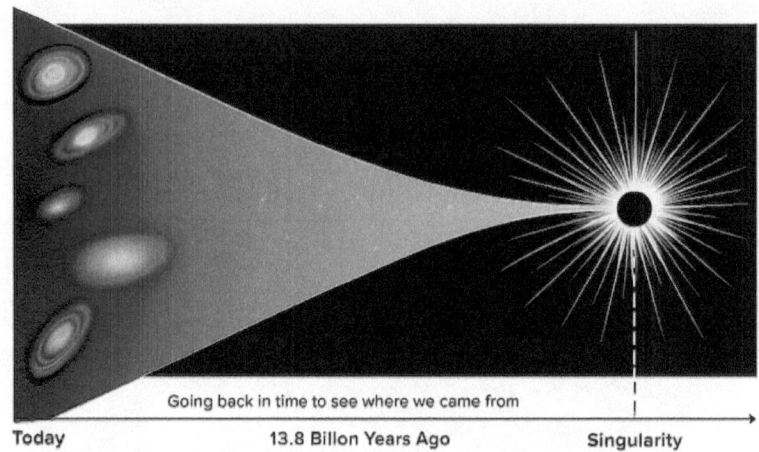

Today   13.8 Billon Years Ago   Singularity

A timelint of the universe. Inoving bealiword from today through 13.3 billion years of cosmic expansion to the singularity where space, time, and energy began.

We do not know how or why the singularity came into existence because science and physics cannot peer at anything before the universe and time and space existed, but in this chapter, I will show how we know that the tiny singularity universe did exist and how its contents made up everything that is in the universe today or ever will be.

The Speed of Light

We will start our story in the 17th century, when a Danish astronomer named Ole Romer made a startling discovery while studying the moons of Jupiter. He noticed the moons appeared late to their predicted positions when Earth was farther from Jupiter. From this he determined that light did not reach him instantly. The further Jupiter's moons were from him the longer it took the light to reach him. He was the first person to discover that light was not instant. It had a speed. Our eyes don't actively reach out to see objects; they passively wait for the light of an object to reach them. Light is extremely fast. It travels at 186,282 miles every second, or 300,000 kilometers per second. It's the fastest thing in the universe, so fast that it just seems instantaneous. But it's not. So, what is light?

What Light Is

Light may look smooth and continuous to our eyes, but at its deepest level it comes in tiny, indivisible packets of energy. Each packet is the smallest "chunk" of energy that light can carry, and physics calls such a packet a quantum. When that quantum belongs to light, it has a special name: a photon. A beam of light is really just an enormous swarm of photons, all traveling together.

Although a photon of light is fast it does not travel in a straight line. It travels in waves. On the way from one point to another a photon will travel as tiny waves that oscillate up and down until they reach their destination, which may be your eyes. How much energy a photon has depends upon the waves it travels in. If the wave is long and drawn-out it is a radio wave and has very little energy, but if it oscillates in very short and fast waves it is a gamma ray and has lots of energy. We can't see either of these waves because our eyes only see a very small and defined section of wavelengths called the visible light spectrum. In the visible light spectrum of photons, we see colors that correspond to wavelengths—longer wavelengths make the visible color red, while shorter wavelengths create the color violet. In between those wavelengths are all the other colors of the rainbow.

Redshift and Blueshift

Another interesting thing about the visible light waves of photons is that they can be stretched or compressed, which changes the color we see. This is called redshift or blueshift. For example, when an object is moving away from you, the light waves reaching your eyes are stretched, so the color appears a little redder than normal —this is redshift. If the object is moving toward you, the waves are compressed, shifting the color toward the shorter wavelengths near blue, a change called blueshift. Technically, the shift can extend into violet, but scientists adopted the term "blueshift" because blue is the more familiar color. These color shifts are far too small to notice with the naked eye, but high-tech instruments can easily detect if an object is moving away or not by the red or blue shift of the object. This principle is important because of what Edwin Hubble discovered in 1929.

The Expanding Universe

That year, Hubble was studying the cosmos and noticed that all the galaxies were redshifted. That means they were moving away from him. Not just some of them, all of them. They were not flying through space like rockets; the galaxies were all moving away from him and away from each other in all directions. Hubble also noticed that the further the galaxies were away from him the faster they were receding. What it meant was that space itself—our entire universe—was expanding. Hubble was able to calculate the rate at which the universe was expanding, which is called the Hubble Constant. This was calculated at approximately 67 kilometers per second per megaparsec (a megaparsec is 3.26 million light years). It is not important that you remember this exact number, as other calculations of universe expansion have slightly different numbers. What is important is to know that the universe is expanding at ever increasing rates.

Now we are getting to the meat of things. If we can use the Hubble Constant to determine how fast the universe is expanding, we can also run the calculations backward over time, like running a film backward, to calculate how quickly the universe would contract. In fact, we can run that film backward with the universe getting smaller and smaller until at some point in time it gets so small there is nothing left to calculate, and all physics breaks down.

That's where there is no space or time—only The Singularity 13.8 billion years ago.

What does that mean? It means that everything in the universe must have originally been shrunk down so small that nothing could have existed. But if nothing could have existed, then how did we get here?

Is There Such a Thing as Nothing?

In 1948, Dutch physicist Hendrick Casimir solved part of this question by proving that there is no such thing as nothing and there never was such a thing as nothing. He proved this with a simple experiment. He put two metal plates very close together in a container and then he removed everything in the container until it contained nothing else—no air, no particles, no light.

And then he waited. Over time, the two metal plates began magically moving toward each other. Casimir concluded that even in a vacuum, like space, there are bits or quanta of energy, enough energy in the vacuum to put pressure on the plates to move them toward each other. What that showed is that there is still energy in space in a place void of everything. This phenomenon is known as the Casimir Effect.

An interesting thing about energy is that it doesn't exist as a substance like matter. It has no mass or volume and therefore does not occupy space. This is called Vacuum Energy, and an infinite amount of vacuum energy can be contained in an infinitely small space. A space like The Singularity.

This is all well and good—we can imagine all the energy of the universe compressed into the Singularity. But energy alone does not explain stars, planets, or people. The missing step is how energy can become matter. That bridge was revealed in 1905 by Albert Einstein, one of the greatest scientific minds in history.

Einstein's Key

Let's look to Albert Einstein for an answer. In 1905 Einstein published his theory of Special Relativity and his famous equation $E = mc^2$, which showed that mass and energy were the same thing in different forms, just like water and ice are the same thing in different forms. This meant that energy could be turned into matter, the matter necessary to build a physical universe. Einstein's theory of Special Relativity contained a startling insight: mass and energy are two forms of the same thing. His famous equation, $E = mc^2$, showed that matter could be transformed into energy and energy into matter, just as water and ice are different forms of the same substance.

This discovery provided the missing link—the mechanism by which the universe's raw energy could be shaped into the matter that would one day form stars, planets, and life itself.

So, we have traced the expansion of the universe from an infinitesimally small point of dense energy that requires no space, and then by adding Einstein's $E = mc^2$ we can turn that massive amount of energy into matter, which creates the physical universe we live in.

Putting It All Together

We use the Hubble Constant to trace the expanding universe back 13.8 billion years to a point too small to even contemplate.

Then we use the Casimir Effect to show that there is vacuum energy in empty space.

Then we show that vacuum energy requires no space, and an infinite amount of vacuum energy can be contained in an infinitesimally small space—like a singularity.

And finally, we use Einstein's equation $E = mc^2$ to turn that energy into matter that does require space.

Looking Ahead

Now comes the big question: how did that tiny point of dense energy transform into the vast ocean of matter that fills our expanding universe? The search for that answer begins in the next chapter.

2 FROM FIRE TO FORM

The creation of a thousand forests is in one acorn.
—Ralph Waldo Emerson
(writer, poet, philosopher)

The Seed of Everything

In the last chapter, we traced the universe back to its mysterious beginning—the singularity. We saw how Hubble's discovery of an expanding cosmos led us to rewind time until space and matter collapsed into a single point, a state where our current physics cannot fully describe what happens. We explored how fields, restless with hidden energy, could exist even in that compressed state, and how Einstein's equation, $E = mc^2$, gave the universe a way to transform that energy into matter. Now we turn to the next step in the story: how the singularity erupted in a burst of expansion called inflation, how the first particles came into being, and how those particles combined to form the earliest atoms.

Inflation—The Universe's Sudden Growth

The singularity was not a point in empty space. It was space itself, compressed into a state smaller than imagination, carrying within it all the energy that would one day become the universe. At that first instant—less than a trillionth of a second after it began—something extraordinary happened. The universe inflated.

Inflation was not a gentle swelling. It was an almost inconceivable surge, a faster-than-light[1] stretching of space itself. Before inflation, the universe may have been smaller than a subatomic particle. Afterward, in less than a heartbeat, it had ballooned to a size far larger than our solar system. This was the growth spurt that set the stage for everything that followed.

The Inflaton Field and the Great Expansion

But what caused this astonishing burst of inflation? The leading idea is a new kind of energy, carried by a field called the inflaton (not misspelled). Just as the electromagnetic field produces photons, the inflaton field carried its own form of energy—one that dominated the early universe.[2]

Before we continue the story of the universe, we need to pause and talk about what a particle is, and the natural laws that every particle must follow.

Particles that emerge from the fields of physics each carry properties that make them unique. In this book we will focus on two of the most basic properties: mass and electric charge.

Mass is the measure of how much matter a particle contains — its "stuff." Importantly, mass does not depend on gravity, so even in empty space a particle's mass stays the same. Protons and electrons, for example, have measurable mass and give matter its weight. Photons, the particles of light, are different: they have no rest mass at all. Yet photons are far from nothing — they carry energy and momentum and can push on matter or bend in gravity's pull.

Electric Charge tells us how particles interact with one another. By protocol, the charge of a particle is written in parentheses after its name—for example, a proton (+1) or an electron (−1). While many familiar particles carry whole-number charges, some particles can have fractional charges. Others, like photons (0), have no charge at all, which means they do not respond to electric forces.

All particles and fields also obey strict rules called the laws of conservation. In this book we will focus on three of these laws: the conservation of energy, momentum, and charge. Each of these laws says that if something changes in one particle, another particle or field must change in a way that keeps the total the same. Nothing is ever lost.

Conservation of Energy - Energy cannot be created or destroyed, only changed. If one particle loses energy, another particle or field must gain the exact same amount. And because of Einstein's equation $E = mc^2$, mass and energy are really the same thing in different forms. Energy can turn into mass (a new particle) or mass can turn back into energy, but the total always balances.

Conservation of Momentum - Momentum is the "oomph" of motion. If one moving particle hits another and stops, the second particle takes on exactly the momentum the first one lost. The total momentum is unchanged.

Conservation of Charge - Charge also balances. If a particle with +1 charge meets a particle with −1 charge, together they make 0. If they annihilate and disappear, the total charge is still 0.

Now, back to our story

In less than the blink of an eye, inflation ended. The inflaton field rapidly decayed, and its energy spilled into the expanding universe, awakening the dormant fields of matter and forces. Among the many fields known to physics, the ones most important to our story are:

1. **Matter Fields**
   a. **Quark Fields**—there are two types of quarks the **Up quark** and the **Down quark** which are the building blocks of protons and neutrons.
   b. **Electron Field**—electrons are matter particles that orbit the protons and neutrons to form atoms.
2. **Force Fields**
   a. **Electromagnetic Field**—which creates photons, the particles of light.
   b. **Strong Force Field**—carried by gluons, which bind quarks tightly inside protons and neutrons.
3. **Higgs Field**—which gives mass to matter particles, allowing them to have weight and structure.

These five fields, along with roughly 20 others described by the Standard Model of Particle Physics[3], filled the universe from the very beginning. During inflation they were stretched across space, lying almost still, quivering faintly in their lowest-energy condition. This dormant state, called the ground state, is the natural resting place of every field.

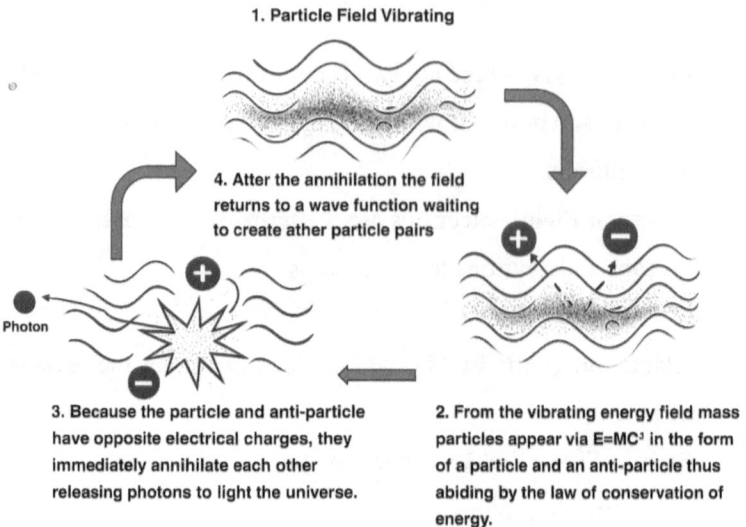

1. Particle Field Vibrating

4. After the annihilation the field returns to a wave function waiting to create ather particle pairs

Photon

3. Because the particle and anti-particle have opposite electrical charges, they immediately annihilate each other releasing photons to light the universe.

2. From the vibrating energy field mass particles appear via E=MC³ in the form of a particle and an anti-particle thus abiding by the law of conservation of energy.

Only when the inflaton decayed—spilling its massive energy into the expanding universe—did the fields awaken. In that instant, the cosmos erupted into the hot chaos of creation, producing trillions upon trillions of fundamental particles. These newborn particles collided furiously at nearly the speed of light, generating unimaginable heat. This moment is what we call the Hot Big Bang.

When the energy surged into the fields, they became excited and produced particle–antiparticle pairs. Each pair balanced the books: opposite charges, opposite momentum, equal energy. Quarks and antiquarks, electrons and positrons, gluons, and photons burst into existence.

Most of these pairs quickly collided with their opposites and annihilated. But the energy did not disappear. Every annihilation released photons—particles of light—in staggering numbers. The early cosmos became a brilliant ocean of radiation, a fireball so intense that light itself could not travel far without crashing into the next particle.

Picture standing in the middle of a forest on a foggy day. You know the trees are out there, but the fog scatters the light in every direction. No matter which way you look, you see only a glowing mist, never the trees themselves. In the early universe, the "fog" was not water droplets, but rather free electrons that were constantly deflecting photons and preventing light from moving in straight lines. The entire cosmos glowed, but nothing could be clearly seen.

And yet, something extraordinary happened. For reasons physicists still do not fully understand, the balance was not perfect. Out of every billion pairs of particles that destroyed one another, one lonely particle survived. This tiny imbalance—just one in a billion —was enough to tip the scales. That slight excess of matter is the reason anything exists at all. Every star, every planet, and every living being is built from those rare survivors of the primordial fire. If more particles had survived, the universe would not be larger, but denser — with more stars, galaxies, and matter packed into the same expanding space.

From those survivors, protons (+1) formed from two up quarks (+⅔, +⅔) and one down quark (−⅓). A neutron (0) formed from two down quarks (−⅓, −⅓) and one up quark (+⅔). Quarks cannot exist alone; they are bound tightly together by gluons (0), the messengers of the strong force, which snap between quarks like invisible elastic bands.

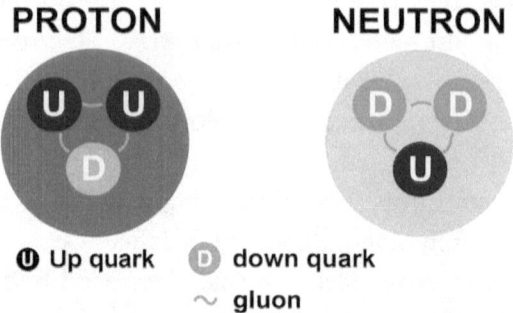

A proton is made of two up quarks and one down quark held together by the strong force of gluons, the neutron is made up of two down quarks and one up quark and which are also held together by gluons.

And yet, charge and force alone are not enough. For these particles to have mass—for them to weigh something and give substance to the universe—they must interact with the Higgs field (o). Without the Higgs field, quarks and electrons would have no mass at all. And protons, neutrons, atoms, and life itself could not exist. The Higgs field is the invisible background that gives particles their heft, turning a universe of pure light into a universe of matter.

Big Bang Nucleosynthesis

At first, the universe was still too hot for particles to stay bound together. Any attempt by protons and neutrons to cling to one another was instantly blasted apart by high-energy radiation. But after about three minutes, the cosmos had cooled to around a billion degrees.

It's worth noting that hydrogen, the simplest element, had already made its first appearance even earlier. A lone proton is itself the **nucleus** — the dense central core — of a hydrogen atom. Because quarks had already joined to form protons within the first second, the universe contained its first hydrogen nuclei almost immediately. But the more complex elements required something new: fusion.

Now, at last, the universe had cooled enough for protons and neutrons to remain bound together instead of being blasted apart. In this calmer state, they could fuse into larger atomic nuclei, the cores that hold nearly all the mass of atoms. This brief but critical phase is called Big Bang Nucleosynthesis. Over the next 20 minutes, some of the free protons fused with neutrons to create helium nuclei and a trace of lithium, while most protons remained as hydrogen.

This early burst of nuclear chemistry set the basic constitution of our cosmos: roughly 75% hydrogen and 25% helium by mass, with everything else to be formed later inside stars.

The Radiant Fog

For the next 380,000 years, the universe remained a hot plasma: a dense soup of free electrons newly formed nuclei made of bound protons and neutrons, and a flood of photons released during billions upon billions of particle/antiparticle collisions and annihilations.

In this plasma, photons could not travel far without crashing into electrons. Light was scattered endlessly in every direction. A single photon might be deflected billions of times each second, never traveling more than a few centimeters before being knocked off course again. The universe glowed, but nothing could be seen through the mist.

Think of it like this, imagine standing in a dense forest on a foggy day. The trees are there, but you cannot see them clearly because countless tiny water droplets scatter the photons of light before they can reach your eyes in a straight line. No matter which way you turn, the view is the same — only a glowing, featureless mist. The early universe was similar, except the "fog" was not made of water droplets but of free electrons, endlessly deflecting photons and preventing light from traveling freely through space.

For 380,000 years the universe shimmered as a boundless sea of light, dazzling yet indistinct, a radiant haze from which all clarity still lay hidden.

A Brief Map of the Road We've Traveled

Inflation smoothed the early universe and set the seeds for structure.

Reheating filled space with energetic particles and radiation.

Asymmetry left a tiny surplus of matter over antimatter.

Quarks and gluons cooled into protons and neutrons.

Big Bang Nucleosynthesis stitched those particles into hydrogen, helium and traces of lithium.

Radiant fog trapped photons of light for 380,000 years.

Each step was not a miracle, but a consequence: cause, effect, and the unhurried patience of physical laws.

The Edge of the New Chapter

The universe had its ingredients, but not yet its clarity. For hundreds of thousands of years, it glowed in a brilliant haze, a fog of light and matter where nothing distinct could emerge. Only when that fog began to thin did the cosmos reveal its first true picture — a silent afterglow that still surrounds us today. In the next chapter, we step into that radiant fog, witness the moment light was set free, and follow gravity as it sculpts the first stars to ignite the darkness.

3 THE RADIANT FOG AND THE FIRST LIGHTS

"We are stardust brought to life,
then empowered by the universe
to figure itself out."
—Neil deGrasse Tyson
(Astrophysicist)

The Brilliant Prison

In the last chapter, we followed the universe from its explosive beginnings to the creation of its first atomic nuclei. But those nuclei, along with free electrons and photons, remained trapped in a seething plasma. Electrons, protons, and helium nuclei whirled chaotically through space, scattering photons at every turn. The cosmos glowed brilliantly, yet no image could pass through the haze because light could not travel far without being deflected.

The Great Cooling - Atoms Formed

For hundreds of thousands of years, the universe remained a hot plasma: a dense soup of free electrons, newly formed nuclei made of bound protons and neutrons, and a flood of photons released during annihilation. Continued expansion caused the temperature to drop to about 3,000 Kelvin ($\approx$ 2,700 °C/4,900 °F), and a quiet turning point slowly arrived roughly 380,000 years after the Big Bang. Electrons finally lost enough energy and slowed down enough to be captured by nuclei, forming the first neutral atoms of hydrogen and helium. This event is called recombination, though it was not a rejoining but the first true union of electrons and nuclei. With free electrons bound away, photons could at last stream freely across the cosmos in straight lines. The fog lifted, and for the first time, light could travel from one end of the universe to the other.

That ancient light is still with us. Through 13.8 billion years of cosmic expansion, the light waves have been stretched (redshifted) from the fierce glow of visible light into the gentle hum of microwaves. Today, we can detect this faint afterglow with special instruments tuned to microwave frequencies.

We call it the Cosmic Microwave Background (CMB)—a ghostly image of when the universe first became visible. It is not just a picture in one direction, but in all directions at once: the oldest light we can ever see, a snapshot of the cosmos just 380,000 years after the beginning.

With the fog lifted and the first atoms formed, the universe changed forever. No longer just a seething plasma of particles and light, it was now filled with neutral hydrogen and helium atoms drifting through space.

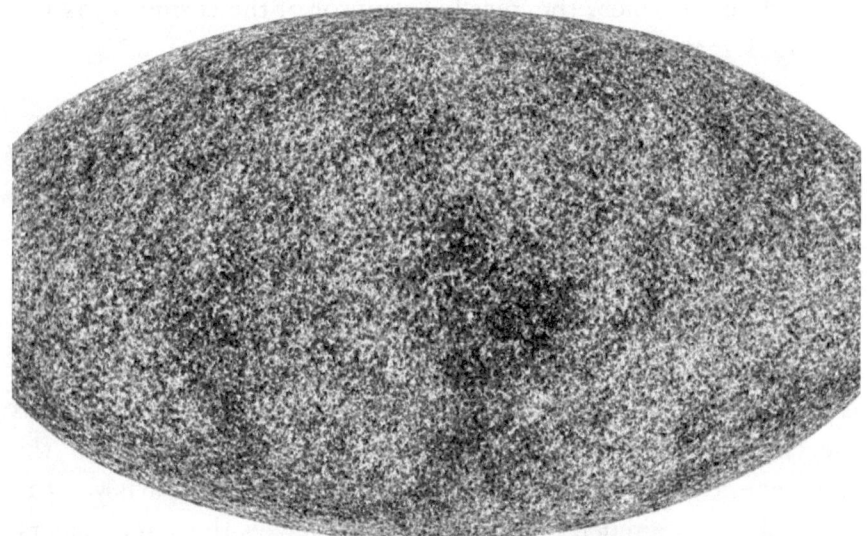

The Cosmic Microwave Background (CMB): Across the immensity of space, every distant light is also a memory. The Sun reaches us eight minutes in the past. The nearest stars in Alpha Centauri shine from four years ago. Andromeda's great spiral arrives from 2.5 million years in our history. The farther we look, the deeper into time we travel. With powerful microwave telescopes, we can follow this trail of ancient light back to a moment 380,000 years after the Big Bang. Before then, the universe was a glowing plasma whose free electrons scattered every photon, keeping the cosmos opaque. When the universe finally cooled, light broke free—and that first light became the Cosmic Microwave Background. Spread across the sky, it is a faint, ancient afterglow. Its smoothness reveals the early universe's remarkable uniformity, while its subtle ripples carry the first seeds of galaxies, stars, and in time, life itself.

The Cosmic Dark Ages

The lifting of the fog did not bring instant brilliance. The universe entered a long, silent adolescence known as the Cosmic Dark Ages. There were still no stars or galaxies—only cooling hydrogen and helium gas, drifting in vast emptiness.

But gravity was already at work.

Tiny density ripples—left over from quantum fluctuations during inflation—meant that some regions contained slightly more matter than others. Where there was more matter, gravity pulled more strongly. Over millions of years, these denser regions attracted more gas, slowly forming great, invisible "wells" of mass.

And here is where the invisible sculptor of the cosmos took the lead: dark matter.

The Dark Matter Mystery

Ordinary matter was not the whole story. Something invisible was also shaping the cosmos. Evidence suggests that dark matter, a mysterious substance that neither emits nor absorbs light, was already at work. Think of a skyscraper: what you see from the street is the brick, mortar, and glass—the facade. Yet hidden inside is a steel framework that does the true structural work and holds everything together. And yet, while the framework for the skyscraper cannot be seen it weighs far more than the parts you can see. Galaxies are much the same. The "facade" is the stars, planets, and glowing gas we can observe, but the real framework is dark matter. And just as the steel in a skyscraper outweighs its facade, dark matter outweighs normal matter by about five to one, providing the hidden gravitational strength that holds galaxies together and gives the universe its unseen structure.

We know dark matter exists because of the way it influences galaxies. When astronomers measured the speeds of stars orbiting within galaxies, they found something surprising: in large spirals like the Milky Way, stars move at about 200–300 kilometers per second (roughly 450,000–670,000 miles per hour). According to the visible matter alone, galaxies spinning that fast should have torn themselves apart long ago. The fact that they remain intact points to a hidden source of gravity. This unseen mass — dark matter — provides the extra pull needed to hold galaxies together. Its presence is also revealed in another way: light from distant galaxies bends more strongly than expected when it passes near massive clusters, betraying the invisible gravity of dark matter. Though it cannot be seen directly, its influence is unmistakable.

Long before hydrogen had gathered itself, dark matter had already clumped into spherical clouds that would one day attract galaxies and galaxy clusters. These vast, unseen scaffolds were scattered throughout the universe and ordinary matter fell into these dark mater clouds guided by their invisible gravity.

Without dark matter's head start, the first stars might have taken far longer to form—or perhaps never formed at all.

For the first few hundred million years after recombination and the cosmic clearing, the universe remained a relatively simple place. Vast clouds of hydrogen and helium drifted through space, slowly gathering under gravity's patient pull. When these clouds grew dense enough and hot enough, they ignited as the first stars—brilliant giants that blazed with the fire of pure hydrogen fusion.

The First Stars—Population III

The universe's first stars, known as *Population III,* were cosmic pioneers—immense, short-lived, and unlike anything we see today. Formed from clouds of pristine hydrogen and helium, they lacked the heavier elements, or "metals," that help modern stars cool and fragment into smaller pieces. With no such cooling agents, these newborn giants grew enormous, often hundreds of times more massive than our Sun.

Their size and density made them burn with extraordinary intensity. Their searing light flooded the cosmos with ultraviolet radiation, splitting atoms apart and reshaping the chemistry of the young universe. But they were fleeting. Many survived only a few million years before their fuel ran out.

Inside their cores, nuclear fusion drove a chain of creation: hydrogen into helium, helium into carbon, carbon into oxygen, oxygen into silicon, and ultimately silicon into iron. At iron, however, the process stalled. Unlike lighter elements, fusing iron consumes energy instead of releasing it. As iron piled up, the balance between the outward push of fusion and the inward pull of gravity collapsed.

The result was spectacular. Some of these stars erupted in supernovae so brilliant they briefly outshone entire galaxies. In those violent explosions, elements heavier than iron—such as gold, platinum, uranium—were forged in an instant, then hurled into space to seed future stars, planets, and eventually life. Others died even more dramatically: the smaller remnants became neutron stars, so dense that a teaspoon of their matter would outweigh mountains, while the largest collapsed into black holes, where gravity crushed everything into singularities—points of infinite density where the known laws of physics break down.

No Population III stars survive today, but their fingerprints remain. We see them in the chemical makeup of younger stars, in the dust of planets, and in every atom of our bodies. Their brief, blazing lives set the stage for all the complexity that followed.

The Second Generation
—Population II

The violent deaths of Population III stars filled space with heavier elements. Over time, gravity pulled these enriched gases back together to form the second generation, or Population II stars.

These stars contained traces of carbon, oxygen, and iron—tiny compared to modern stars, but enough to change everything. Metals allowed gas clouds to cool more efficiently, enabling smaller, longer-lived stars to form.

Many Population II stars still shine today, especially in globular clusters—spherical swarms of ancient stars orbiting the cores of galaxies. They are among the oldest surviving witnesses of the universe, holding chemical records of an era billions of years past.

While less massive than most Population III stars, some extremely massive Population II stars can still die as supernovae, further enriching the cosmic medium with heavier elements. With each cycle of birth and death, the chemical diversity of the universe increased.

Kilonova—The Universe's Rarest Forge

Among the most extraordinary events in the cosmos are kilonovae —cataclysmic collisions between neutron stars.

Picture two neutron stars, each the compressed remnant of a supernova, locked in a gravitational embrace. Over millions or billions of years, they spiral slowly inward, radiating energy as gravitational waves—ripples in the fabric of space-time. At the end, they merge in a moment of staggering violence.

A kilonova releases more energy in seconds than our Sun will produce over its entire lifetime. The impact shakes the universe, sending both light and gravitational waves across billions of light-years.

But its true wonder lies in its creative power. In the unimaginable heat and pressure of a kilonova, the heaviest elements in the periodic table are forged — elements even supernovae cannot make in abundance. Gold, platinum, thorium—precious not just for their rarity, but for the extraordinary conditions required to create them —are born in these brief moments.

Every gold ring, every silver coin, every uranium atom on Earth exists because somewhere, long before our Sun was born, two dead stars collided.

The Third Generation—Population I

After countless cycles of stellar birth and death, the universe reached a tipping point. Gas clouds had become rich with metals, enabling the formation of stars like our Sun—Population I stars.

These younger stars are more chemically complex, with enough carbon for life, oxygen for water, silicon for rock, and iron for planetary cores and magnetic fields. They form in the spiral arms of galaxies, often surrounded by disks of gas and dust—the raw materials for planets.

Our Sun is one such star, born about 4.6 billion years ago from a metal-rich cloud enriched by generations of stellar ancestors. The Earth—and all life upon it—emerged from this long chain of cosmic alchemy. None of it would have been possible had there not been just a tiny excess of matter over antimatter in the early universe: for every billion particles of antimatter, there was one extra particle of matter left behind. That minuscule imbalance tipped the scales, allowing stars, planets, and eventually life to form. From that razor-thin margin, the universe has continued to grow ever more complex—atoms into stars, stars into worlds, and worlds into living beings capable of telling the story

A Universe Forged in Struggle

The universe did not begin ready for life. It had to build the necessary elements in stages, through the lives and deaths of stars. Every atom of oxygen in your lungs, every calcium atom in your bones, every trace of gold in a wedding band—was created in the heart of a star or in the violence of a cosmic collision.

We are not the product of a single star's death, but of many. Our existence is the most recent verse in a song the universe has been composing for over 13 billion years.

Coming Up Next

In the next chapter, we turn from the grand stage of the universe to a more intimate theater—the birth of solar systems. We'll see how gravity sculpts not just stars, but worlds, and how, around one particular star, the chemistry of life found its first home.

From stardust, biology.

From chaos, chemistry.

From fire and fog, the blueprint of life.

The journey continues.

4 THE SOLAR SYSTEM AWAKENS

*"We are children of ancient embers,
born from fire, sustained by sunlight,
destined to return to the stars."*
—Anonymous proverb

Children of Ancient Embers

In the last chapter, we watched the universe emerge from its long adolescence. The fog lifted, light was set free, and the first stars ignited giants that lived short, brilliant lives before exploding into supernovae or collapsing into black holes. With every stellar death, new elements were scattered like seeds into space. Gold, oxygen, carbon, silicon—all the raw materials of future worlds—were forged in those furnaces of creation.

Yet this was only the beginning of the universe's chemistry. For hundreds of millions of years, stars were born and died, enriching the cosmos with each cycle. What had begun as a simple universe of hydrogen and helium became steadily more complex, a laboratory where matter learned new forms. Out of this slow alchemy, galaxies grew rich in the ingredients for planets, oceans, and eventually life itself.

Now we come to the next great turning point: the birth of our own solar system. Around 4.6 billion years ago, in one corner of the Milky Way, the ashes of countless ancient stars gathered once more. Gravity pressed them into a swirling disk of gas and dust, where a middle-aged star—our Sun—flared into being. From that same disk, rocky planets, gas giants, and icy worlds began their dance. Among them was a small, stony sphere that would one day host seas, continents, and the first stirrings of life.

This is the story of how death made life possible—how the remains of ancient suns became the Earth beneath our feet and the spark of biology within us.

The Birth of a Star

Fast-forward to about 4.6 billion years ago in a quiet spiral arm of the Milky Way. In a cloud of gas and dust enriched by countless generations of stellar birth and death, something stirred.

Perhaps it was the shock wave from a nearby supernova, or the gravitational tug of a passing star, or the inevitable instability of a cloud too massive to support itself. Whatever the trigger, a molecular cloud began to collapse under its own gravity.

As it contracted, it spun faster—like a figure skater pulling in her arms—and flattened into a disk. The center grew hotter and denser, pressure building until the core reached about 10 million •Kelvin (18 million •F). Then the spark caught: hydrogen nuclei fused into helium.

Our Sun was born.

Around it spun a pancake-shaped protoplanetary disk, a construction yard of rock, metal, ice, and gas—debris from dead stars ready to assemble into worlds.

The Architecture of Worlds

Inside this disk, physics sorted matter by temperature and distance from the newborn Sun. Close to the star, heat stripped away volatile materials, leaving behind only metals and rocky minerals—destined to become the inner, terrestrial planets. Farther out lay the frost line—the invisible boundary where temperatures were low enough for water, methane, and other compounds to freeze into ice. Beyond it, planetary cores could grow larger and massive enough to capture thick atmospheres, becoming the gas giants. Tiny dust grains stuck together through static cling, growing into pebbles, then boulders, then planetesimals—ranging from mountain-sized to nearly planetary. Collisions could be violent, shattering some bodies into fragments that became asteroids, while others merged to form the planets we know today.

Earth's Violent Birth and the Theia Impact

Earth's formation was anything but peaceful. The young planet, still molten from relentless asteroid and comet bombardment, orbited in a high-stakes demolition derby. But the most dramatic event came about 4.5 billion years ago.

A Mars-sized protoplanet—now known as *Theia*—was on a crossing orbit with the early Earth. Over millions of years, gravitational nudges drew the two worlds onto a collision course. When they met, the impact unleashed energy equivalent to tens of billions of nuclear bombs.

Theia's iron core merged with Earth's, helping give our planet a dense metallic heart and a powerful magnetic field. Much of Theia's rocky mantle was blasted into orbit as molten debris. From a few months to several years, gravity sculpted that debris into a single large body—our Moon.

The collision reshaped Earth's destiny. It tilted our axis about 23.5°, giving us seasons. It likely stripped away an early atmosphere, paving the way for the one we now breathe. The added spin from the impact shortened Earth's day, speeding its rotation. Most importantly, the Moon's steadying gravitational influence has kept Earth's climate relatively stable for billions of years—critical for life's eventual rise.

Without Theia, Earth might have been a wildly wobbling planet with extreme climate swings, perhaps too unstable for complex life to persist.

A World Finds Its Balance

In the wake of the Moon-forming impact, Earth remained a dangerous place—its surface a molten ocean of magma, its skies thick with steam and volcanic gases. But slowly, the planet began to cool.

The crust solidified into the first fragile continents. Water vapor condensed, and rains—possibly lasting thousands of years—filled basins to form the first oceans. Lightning cracked across dark skies, igniting chemical reactions in the new seas.

Between 4.1 and 3.8 billion years ago, Earth endured a violent chapter known as the Late Heavy Bombardment. During this period, gravitational shifts among the giant planets sent swarms of asteroids and comets hurtling into the inner solar system. Our young planet was pummeled relentlessly: giant impacts could vaporize entire oceans in an instant, while smaller collisions stirred the atmosphere and delivered fresh supplies of water, carbon compounds, and metals from the outer solar system. Some scientists believe this rain of material was crucial in providing the raw ingredients from which life would eventually emerge. As the giant planets settled into stable orbits, the storm of stray asteroids and comets began to subside. Most were either swept up in collisions, flung out of the solar system, or confined to calmer belts of debris. With the barrage easing, Earth's surface stabilized. Oceans persisted. The atmosphere thickened with nitrogen, carbon dioxide, and water vapor. Volcanoes enriched the air and seas with minerals. Tidal forces from the Moon stirred the coasts, creating nutrient-rich shallows—ideal laboratories for prebiotic chemistry. Earth, once a molten casualty of celestial violence, had found a balance between chaos and stability—a balance that would cradle the first living systems.

The Celestial Dance

While Earth was settling, the rest of the solar system was finding its own equilibrium. Jupiter and Saturn—gravitational giants—played a critical role in shaping the solar system's architecture.

Early in the solar system's history, Jupiter and Saturn did not sit where they do today. As they interacted with the surrounding protoplanetary disk of gas and dust, their immense gravity caused them to drift inward toward the Sun. Jupiter may have moved as close as Mars's present orbit before reversing course. When Saturn grew large enough, it too was drawn inward, and the two giants became locked in a gravitational resonance. This pairing changed how they tugged on the gas disk, flipping the balance of forces and pushing both planets back outward—though always under the Sun's dominant gravity.

This unusual inward–outward journey is called the Grand Tack, borrowing a term from sailing. A "tack" is when a sailboat changes direction by turning its bow through the wind; Jupiter, like a cosmic ship, first steered inward and then abruptly turned back outward with Saturn as its partner. The consequences of this migration were profound. Jupiter's inward sweep scattered much of the material that might have built a larger Mars, leaving the Red Planet small. On its way back out, Jupiter also stirred up the asteroid belt, mixing rocky bodies from the inner solar system with icy ones from farther out — a blend that still survives in the belt today.

Even now, gravitational resonances ripple through the system. Jupiter's pull flexes the interior of its moon Europa, keeping its subsurface ocean liquid. Saturn's influence heats Enceladus, driving icy geysers that vent into space. The giant planets also act as cosmic bodyguards, deflecting some comets and asteroids that might otherwise strike Earth.

This is not a static system but a dynamic ecosystem—an interlinked choreography where planets, moons, and debris constantly influence one another over millions of years.

The Solar System as Ecosystem

The solar system emerged not as a collection of isolated worlds but as an interdependent web. Comets deliver water and organics. Asteroids occasionally reset evolutionary clocks with extinction events—only for life to rebound in new forms. The Sun's core slowly contracts as it fuses hydrogen into helium, raising its temperature and causing fusion to run faster. As a result, the Sun shines a little brighter with each passing eon, gradually altering conditions on every planet. In a billion years, Earth's oceans will begin to boil. In about five billion, the Sun will swell into a red giant, likely engulfing Mercury and Venus, and possibly Earth. But here in the Sun's middle age, stability reigns—long enough for life to emerge, evolve, and look back on its origins.

A Meditation on Home

Standing on Earth today, watching the Sun set over landscapes shaped by billions of years of geology, we are witnessing the culmination of an extraordinary cosmic story. The ground beneath our feet is made of elements forged in ancient stars. The water in our veins may have arrived on comets four billion years ago. The air we breathe was shaped by the earliest life forms.

We are not observers from the outside—we are the solar system made conscious. From stellar death came planetary birth. From chaos came stability. And from the ashes of ancient suns came a world capable of knowing itself.

But the story was far from over. For all its beauty, a lifeless Earth would still be only a silent stone adrift in space. In those restless seas, something wondrous was stirring — for the first time here, and perhaps anywhere. From simple molecules came the first trembling steps of self-replication, a fragile spark cradled in water and chance. Out of that spark — so delicate it might have flickered out at any moment — emerged the first stirrings of life, the seed from which every leaf, every fin, and every thought would one day grow.

5 THE SPARK BEFORE LIFE—MOLECULES BEGIN TO ORGANIZE

"The origin of life must be a highly probable affair...given the initial conditions, life will emerge."
— **Carl Sagan**

The Dancing Atoms

In the last chapter, we followed how dying stars scattered their atomic ashes across the galaxy—carbon, oxygen, silicon, iron, and dozens of other elements that would one day build worlds. We watched Earth emerge from cosmic chaos as a molten planet slowly cooled, water gathered in primordial oceans, and volcanoes hissed into a sky thick with strange gases. Thunderclouds rolled across that dim, storm-choked world; lightning split the gloom while the young, fierce Sun poured radiation onto a restless surface where chemistry was waiting to begin its most extraordinary experiment. But nothing lived. Not yet.

No cells pulsing with purpose. No instincts driving survival. No memories encoding the past. Just atoms—those stellar ashes—

dancing in wild, seemingly random patterns across a planet that knew neither life nor death, only transformation.

The oceans churned with heat and minerals dissolved from newborn rocks. The sky flickered with electrical storms that dwarfed anything we see today. If a conscious observer had been watching from some distant vantage point, it might have seen only chaos—endless, purposeless chemical reactions in a world without meaning.

But nestled within that apparent chaos were the seeds of everything to come. Every laugh, every tear, every moment of wonder you have ever experienced was quietly assembling itself in those ancient seas as something unprecedented happened: atoms began to combine with one another to form larger structures that could persist, grow, and eventually remember.

From Atoms to Molecules

These atoms—the scattered remnants of ancient supernovae—arrived on Earth as individuals, floating free in the oceans and atmosphere, each one a single unit with its own properties.

But atoms, it turns out, could be categorized as social creatures. They have a powerful tendency to reach out to other atoms, forming bonds that create entirely new structures with entirely new properties. When atoms link together in specific patterns, they create molecules—and molecules are where chemistry gets truly interesting.

Molecules: The First Molecular Architects

A molecule is a tiny architectural structure built from atoms—but it's more than just the sum of its parts. Think of atoms as universal building blocks, like LEGO bricks that can snap together in countless ways according to the invisible rules of electrical attraction and repulsion. A molecule is what emerges when you connect these atomic pieces, creating structures with properties that neither individual atom possessed alone.

Consider water: take two hydrogen atoms and one oxygen atom, link them together in just the right way, and suddenly you have something completely different—a liquid that can flow, freeze, boil, and support life. The hydrogen atoms alone are explosive gas. The oxygen atom alone supports combustion. But together, as a water molecule, they become the solvent of life itself.

Water is a molecule: two hydrogen atoms embracing one oxygen atom. Carbon dioxide is a molecule: one carbon flanked by two oxygen atoms. So is sugar, so is salt, so is the complex protein that allows you to read these words. Everything you see, touch, breathe, or eat is made of atoms—including every molecule in your body, every thought in your mind.

But early Earth's molecules were relatively simple—molecular beginners in a cosmic chemistry class. They drifted through the oceans and floated in the atmosphere, bumping into each other in what seemed like random encounters. Sometimes they would stick together briefly, then break apart. Sometimes they would react violently, releasing energy in flashes of heat and light.

It was chaos, yes—but a creative kind of chaos. The kind that builds worlds.

What makes this even more remarkable is that these molecular encounters were not truly random. They followed rules—the invisible laws of chemistry and physics that govern how atoms attract, repel, and combine. Even in apparent chaos, order was quietly asserting itself.

Earth as a Chemistry Laboratory

Picture early Earth as nature's most ambitious chemistry experiment—a laboratory the size of a planet, running countless reactions simultaneously for hundreds of millions of years.

The surface was violent beyond imagination. In addition to

erupting volcanoes and incredible lightning storms, ultraviolet radiation from the young sun streamed down through an atmosphere that lacked the protective ozone layer [1]we depend on today. Asteroids and comets regularly slammed into the surface, each impact releasing energy equivalent to millions of nuclear bombs.

And beneath all this chaos, the oceans churned with heat and minerals—a primordial soup rich with the raw ingredients of complexity.

This planet was conducting chemistry on a scale never seen in the solar system. Simple atoms like carbon (C), hydrogen (H), nitrogen (N), oxygen (O), phosphorus (P), and sulfur (S) combined under the relentless pressure of heat and energy. The very violence that might have seemed destructive was actually creative, forcing atoms into new arrangements, new partnerships, new possibilities.

Comets arriving from the outer solar system may have delivered even more complex molecules, including carbon-based compounds forged in the cold depths of interstellar space. These cosmic deliveries added exotic ingredients to Earth's expanding chemical repertoire.

From millions of years of these chaotic interactions, four key types of organic molecules began to emerge with increasing frequency:

1. Amino acids—the building blocks that would one day construct proteins

2. Simple sugars—future sources of energy and structural materials

3. Fatty acids—molecules with split personalities, part water-loving, part water-fearing

4. Nucleotides—the molecular "letters" that would eventually spell out the code of life

These weren't products of design or intention. They were the inevitable result of chemistry following its own internal logic across deep time.

Lightning That Made Life

In the 1950s, two scientists named Stanley Miller and Harold Urey decided to recreate early Earth conditions in their laboratory. It was an audacious experiment—attempting to compress hundreds of millions of years of planetary chemistry into a simple glass apparatus.

They filled a sealed flask with gases thought to resemble Earth's early atmosphere: methane, ammonia, water vapor, and hydrogen. Then they zapped this mixture with electric sparks, simulating the lightning storms that constantly illuminated the ancient world.

The results were startling. After just a few days, the clear liquid inside their flask had turned pink, then brown, then a rich reddish color. When they analyzed this brew, they found it teeming with amino acids—the fundamental building blocks that life uses to construct proteins.

This was a profound revelation. Life's ingredients could form naturally, spontaneously, without magic or divine intervention—just from energy and chemistry following their own inherent patterns. The molecules of life weren't special exceptions to natural law; they were natural law expressing itself through chemistry.

The Miller-Urey experiment suggested something both humbling and magnificent: given the right conditions and enough time, the universe seems almost eager to generate the building blocks of life. Chemistry, it appeared, has a bias toward complexity.

How Molecules Learned to Self-Organize

Among the molecules forming in Earth's ancient oceans, fatty acids played a particularly crucial role—and they demonstrated something remarkable about how order can emerge from chaos.

Fatty acids are molecules with split personalities. One end of these molecules loves water and dissolves readily in it. The other end hates water and avoids it whenever possible.

When you place these schizophrenic molecules in water, something magical happens. They don't just float around randomly. They spontaneously organize themselves. Their water-loving tails reach toward the water, while the water-hating nodes cluster together, hiding from their aquatic environment. The result? They naturally form hollow spheres—perfect little bubbles floating in the ocean.

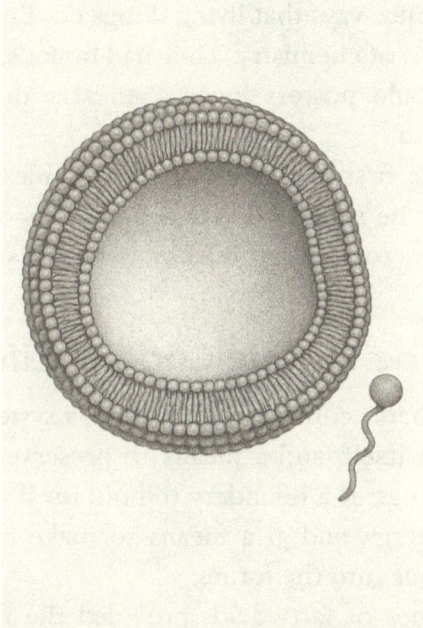

A primitive protocell with its lipid bilayer membrane. The spherical fatty acid molecule on the right shows the hydrophilic (water-attracting) head and hydrophobic (water-repelling) tail, the basic structure that allowed membranes to self-assemble and create the first cellular boundaries.

But these hollow balls of bunched-together fatty acids weren't just any bubbles. They were the first membranes—thin but protective barriers that could surround and contain other molecules. They created boundaries between inside and outside, between self and environment. They could trap useful chemicals inside while keeping harmful ones out.

Imagine them as microscopic water bubbles drifting through ancient seas, each one carrying its own tiny sample of the primordial chemical soup. Some bubbles might contain more interesting molecules than others. Some might be more stable, lasting longer before breaking apart.

We call these structures protocells. They weren't alive—not yet. They couldn't reproduce themselves or respond to their environment in the complex ways that living things do. But they were more than random blobs of chemistry. They had boundaries, they showed behavior, they could possess inner chemistry different from the ocean around them.

They were the first test tubes or test bubbles, where different chemistries could be tested billions and trillions of times a second for billions of years to see what works and survives.

The Simple Code of Life

But life is more than a container. To endure, a system must also find energy to sustain itself, and a means to preserve its pattern. Life requires three things: 1) a boundary to hold itself together, 2) a way to draw upon energy, and 3) a means to make copies so that its design can continue into the future.

The membranes of fatty acids provided the first step. Within them, Earth's elements began their endless trials, combining and recombining in every possible way. At the time, about 16 elements were available in Earth's oceans and atmosphere to take part in these experiments. Yet not all of them could rise to the task. Some formed only fleeting unions. Others were too scarce, too unruly, or too inert. The natural laws acted as a sieve, and through that sieve only a handful of elements passed. Four elements would become life's foundation: carbon, hydrogen, oxygen, and nitrogen.

Carbon, the great architect, able to connect in four directions and build chains, rings, and branching frameworks.

Hydrogen, the simplest atom, everywhere and eager to join in.

Oxygen, the magnet, bringing polarity and allowing molecules to clasp and release each other with precision.

Nitrogen, subtle and versatile, shaping flat, stable structures that could line up in repeating patterns.

Together they formed a quartet with extraordinary talents: the ability to build, to interact, to store energy, and to encode information. From these four elements, chemistry could spin out an almost endless variety—billions upon billions of possible molecules, each one like a ticket in a cosmic lottery. For ages beyond imagination, the oceans of Earth played out this lottery, shuffling and reshuffling combinations. Most combinations failed, vanishing as quickly as they formed. But across billions of years of trial and error, a handful of molecules endured, proving stable enough and subtle enough to carry instructions. Four of them—adenine, uracil, cytosine, and guanine—were the winners.

These were the nucleotides, the first letters in life's alphabet, and the foundation of a code that would change the planet forever.

The First Letters of Life—Nucleotides

With the bases in place—adenine, uracil, cytosine, and guanine—the chemistry of Earth had discovered its alphabet. Each base was a special kind of molecule, built mostly from carbon and nitrogen, like letters waiting to be written. On their own they were scattered characters, but when each one snapped onto a sugar, it found its place on the string. The sugars acted as connectors, holding the letters in line, while phosphates served as the knots that tied one sugar to the next.

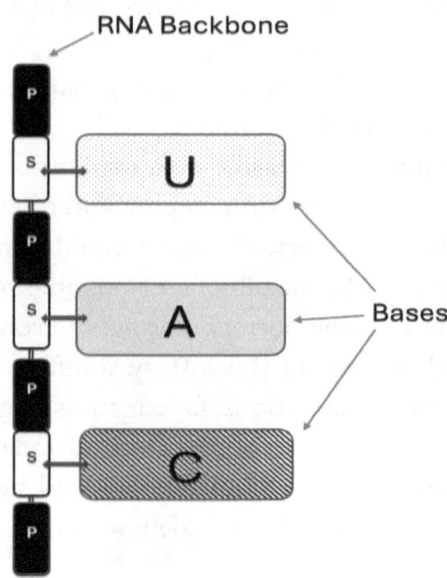

RNA Backbone

Bases

Each unit shown here is a
nucleotide, made of three parts:
Base (the letter: A, U, C, or G)
– carries information, like a
letter in life's alphabet.
Sugar (S, ribose) – the
connector, holding the base
on one side and linking to the
phosphate on the other to
form the chain.
Phosphate (P) – the acidic glue
that links sugars together,
building the backbone of
RNA.
Together, bases, sugars, and
phosphates create the repeating
structure of nucleic acids.

Together they formed nucleotides—small three-part units that could link into long ribbons or backbones. And along those backbones, patterns appeared: sequences that could be read, remembered, and copied. What had once been blind chemistry now carried the first hints of a message. The nucleotides were more than molecules; they were letters, and together they would write the opening lines of life's story.

At first, these chains formed randomly, broke apart, and formed again in different combinations. Most of them were probably quite boring—just sequences of chemical letters with no particular meaning or function.

But occasionally—very rarely at first—one of these molecular necklaces would fold into something special.

RNA—The Molecule That Copied Itself

Among all the nucleotide chains forming in the ancient oceans, one type proved revolutionary: RNA, or ribonucleic acid. Unlike simple molecular chains that just dangled in the water, RNA had a remarkable ability—it could fold back on itself, creating complex three-dimensional shapes.

Each fold created loops, curls, and bends, and some of these geometric arrangements had extraordinary properties. A few folded RNA molecules could grab other molecules from the surrounding water and hold them in precise positions. Others could speed up chemical reactions that would otherwise occur very slowly. Still others could break apart molecules or help build new ones.

These shape-shifting RNA molecules were called ribozymes— RNA enzymes that could do work. Think of them as molecular tools: some like wrenches that could twist other molecules, others like magnets that could attract and organize raw materials, still others like scissors that could cut chemical bonds with precision.

But the most remarkable ribozymes of all were those that could help build new RNA strands. They could attract the right nucleotide beads from their environment and link them together in specific patterns. The process was crude and error-prone, but it worked. For the first time in Earth's history, molecules had gained the ability to make copies of themselves. And though this was still not life, it was revolutionary.

An RNA strand that could copy itself, even imperfectly, had a tremendous advantage over molecules that could only form through random chemical reactions. Over time, better-copying, more stable RNA strands began to outnumber their less capable cousins.

This was no longer random chemistry. This was selection—the first whispers of evolution, driven not by survival in the biological sense, but by persistence in the chemical realm.

The RNA World— Chemistry's First Memory

Scientists now believe there was a long period in Earth's early history—perhaps hundreds of millions of years—when RNA molecules were the closest thing to life on the planet. This era is called the RNA World, and it represents one of the most important chapters in the story of how chemistry became biology.

In the RNA World, these remarkable molecules served dual roles that no modern organism attempts: they were both library and librarian, both database and computer. They could store information in their sequences of nucleotide letters, and they could also do work—catalyzing reactions, building structures, making copies of themselves and others.

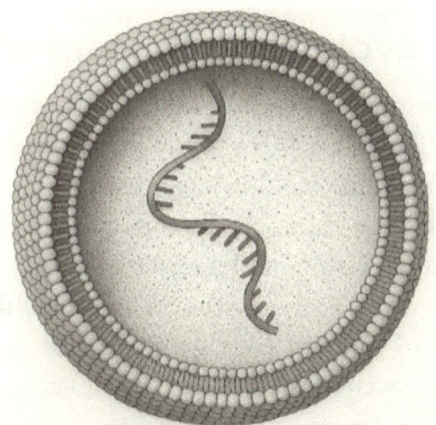

A model of an early protocell. The outer layer shows a
membrane made of fatty acids, forming a protective bubble
that separates inside from outside. Within it lies a strand of
RNA, built from bases (adenine, cytosine, guanine, and uracil)
attached to a sugar–phosphate backbone. The membrane
provided the **boundary**, while RNA carried the **information**
—together representing two of the essential requirements for
the first life.

Picture oceans filled with protocells, each one containing its
own collection of RNA molecules. The protocells with the most
helpful RNA would tend to grow larger, last longer, and eventually
split into two smaller protocells—each carrying copies of the
successful RNA recipes.

This RNA World wasn't quite life as we know it, but it was
getting tantalizingly close. For the first time, Earth had chemical
systems that could store information, pass that information to the
next generation, and even occasionally improve through beneficial
errors.

But RNA had a critical weakness: it was fragile. RNA molecules
could break apart easily when exposed to heat, ultraviolet radiation,
or certain chemicals. A library written in RNA was constantly at
risk of losing its most important volumes.

Nature needed a more durable way to store information—a
more stable molecular memory.

DNA—Evolution's Backup System

The solution that emerged was elegant: DNA, or deoxyribonucleic acid. DNA was like RNA in its basic structure, but it made one crucial improvement that would change everything.

Instead of existing as single strands that folded into shapes, DNA came in pairs. Two complementary strands wrapped around each other in a graceful spiral—the famous double helix that has become the icon of modern biology. The nucleotide bases formed the "rungs" of this twisted ladder: A (adenine) always pairing with T (thymine), and C (cytosine) always pairing with G (guanine). Note that DNA replaced RNA's uracil (U) with thymine (T), a slightly more stable alternative, which makes it easier to spot and fix errors that would be missed if uracil were used instead.

This double-stranded design was brilliant for information storage. If one strand became damaged, the other strand could serve as a template for repair. The code was now backed up, redundantly stored, protected against the chemical chaos of early Earth.

DNA didn't fold into catalytic shapes the way RNA did. Instead, it specialized in being a molecular library—a stable, long-term repository for information. But what kind of information?

Think of DNA as containing thousands of different "recipes"— each one a precise set of instructions for building a specific protein. These molecular recipes are called *genes*. Just as a cookbook contains many different recipes for different dishes, DNA contains many different genes for different proteins.

Each gene is like a sentence written in the four-letter alphabet of DNA (A, T, C, G). Some genes are short sentences of a few hundred letters. Others are long paragraphs stretching across thousands of letters. But each one spells out exactly how to construct one particular protein, step by step.

But how does the cell know where one gene ends and another begins? DNA contains special "punctuation marks"—specific sequences of letters that act like periods and capital letters in written language. These molecular punctuation marks tell the cell's machinery: "Start reading here" and "Stop reading here." Without this punctuation, genes would run together into meaningless gibberish.

DNA was the archive, not the factory—a vast library of genetic gene recipes that could be safely stored and precisely copied.

But DNA couldn't work alone.

The Great Partnership: DNA, RNA, and Proteins

As DNA emerged as the preferred storage medium for genetic information, life stumbled upon one of its most important innovations: the division of labor.

DNA became the master library, safely storing the complete genetic instructions.

RNA evolved into the messenger and translator, copying specific sections of the stored information and carrying them to where they were needed.

And a new player entered the game: *proteins*. These were large, complex molecules built from chains of amino acids. Unlike RNA, proteins couldn't store genetic information or copy themselves. But they could fold into incredibly intricate three-dimensional shapes that performed highly specialized functions.

Proteins became life's workforce. Some proteins became enzymes that could speed up chemical reactions with extraordinary precision. Others became structural materials, building the scaffolding of cells. Still others became molecular motors, transport vehicles, or communication devices.

This trio—DNA, RNA, and proteins—formed the foundation of what we now call the Central Dogma of molecular biology: information flows from DNA to RNA to proteins. DNA stores the plans, RNA carries the messages, proteins do the work.

And here's the crucial insight: each component of this system depends completely on the others. DNA cannot build proteins without RNA to serve as messenger. RNA cannot be manufactured without proteins to do the construction work. Proteins cannot be created without DNA to provide the blueprints.

This circular dependency created something unprecedented: a

self-sustaining, self-repairing chemical system. It was a feedback loop that could maintain itself, improve itself, and most remarkably, reproduce itself.

Chemistry had crossed an invisible threshold. It had become biology.

Then what is life?

When exactly did molecules become "life"? This is one of the most profound questions in science, and the honest answer is that there was no single magical moment. Life emerged through a gradual process, a series of chemical innovations that accumulated over vast stretches of time.

First, molecules gained memory—RNA strands that could store information and copy themselves, however imperfectly.

Then, they gained boundaries—fatty acid membranes that created distinct cellular spaces, separating internal chemistry from the outside world.

Finally, they gained coordination—DNA storing stable genetic plans, RNA carrying those plans to the construction sites, and proteins executing the plans with molecular precision.

Life is not one thing. It is a system—an intricate web of chemical processes that can build itself, maintain itself, reproduce itself, and even improve itself through small, random changes that occasionally prove beneficial.

This is perhaps the most remarkable insight from studying life's origins: life is not a substance or a special ingredient. It is an organization, a pattern of chemical relationships that creates something greater than the sum of its molecular parts.

Mutation—Life's
Engine of Innovation

Once the DNA-RNA-protein system was established, something beautiful happened: copying errors became a source of creativity.

Each time DNA made a copy of itself—whether during cell division or repair processes—small mistakes inevitably crept in. A nucleotide base might be switched for a different one or occasionally added or deleted from the sequence. Most of these mutations were neutral or harmful, disrupting important proteins or regulatory sequences.

But occasionally—very rarely—a mutation would improve something. It might make a protein work more efficiently, or help a cell survive in a particular environment, or allow a membrane to remain stable under stressful conditions.

And those tiny advantages would be passed on to the next generation of cells.

This was evolution in its purest form: not guided by intelligence or intention but shaped by the simple filter of survival. What worked persisted and spread. What didn't work disappeared.

Mutation provided the raw material for evolution's endless creativity, while natural selection provided the editorial process that determined which innovations would survive.

The First True Cells

Eventually, after countless generations of chemical evolution, some protocells achieved the full integration of all life's essential systems:

1. DNA molecules that stored complete genetic libraries—collections of genes, each encoding the instructions for building a specific protein

2. RNA molecules that could read those genes and carry their messages to the protein-construction machinery

3. Proteins that could perform thousands of specialized functions—from metabolism to membrane maintenance to DNA repair

4. Cellular membranes that created controlled internal environments and regulated the flow of materials in and out of the cell

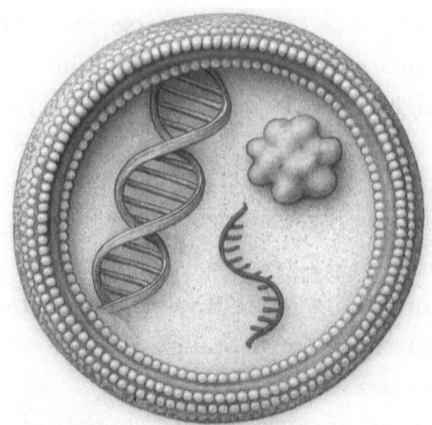

An illustration of a protocell with a fatty acid membrane enclosing the key molecules of life. Inside are strands of **DNA** (the long double helix), **RNA** (the shorter single helix), and a folded **protein** (the tangled black structure). Together, these represent the three central players of biology: DNA storing information, RNA carrying and relaying instructions, and proteins carrying out the work of the cell.

These weren't just organic blobs floating in the ocean anymore. They were cells—integrated, self-sufficient units of life. Not merely chemistry following its own patterns, but biology pursuing its own agenda.

Each cell was a microscopic world unto itself, containing all the molecular machinery needed to grow, survive, reproduce, and evolve. They were chemical systems that had learned to be individuals.

And once they began dividing—one cell becoming two, two becoming four, four becoming countless millions—nothing on Earth would ever be the same again.

The planet had crossed a threshold from which there was no return. Chemistry had become biology, and biology was about to remake the world.

Closing Reflection—The Spark That Changed Everything

Life didn't begin with a flash of lightning or a moment of divine inspiration. It emerged through a flicker—molecules that slowly learned to bend, bond, and most importantly, replicate their successful arrangements.

When chemistry discovered memory, replication, and the ability to improve through beneficial errors, it crossed an invisible threshold that separated the living from the non-living. The spark of life wasn't a miracle—it was the inevitable consequence of chemistry following its own deepest patterns across hundreds of millions of years.

What seems almost miraculous from our perspective is how chemistry, given enough time and the right conditions, seems to naturally evolve toward greater complexity and organization.

We as people are the inheritors of that ancient chemical breakthrough—temporary arrangements of the same atoms that once floated freely in primordial seas, now organized into patterns complex enough to ponder their own origins.

Looking Ahead—The Long Reign of Microbes

The first cells were structurally simple compared to modern organisms, but they were persistent beyond measure. For nearly a billion years, they would rule the planet—alone, unseen by any eyes, utterly dominant in every ocean, lake, and puddle on Earth.

In the next chapter, we'll follow their quiet but revolutionary impact on the planet itself. We'll see how these microscopic pioneers transformed the atmosphere, poisoned the oceans with a deadly new gas called oxygen, and set the stage for every complex form of life that would follow—including plants, animals, and eventually humans capable of understanding their own remarkable origins.

The age of chemistry was ending. The age of biology had begun. And biology was about to change everything.

6 THE OXYGEN REVOLUTION —WHEN LIFE LEARNED TO BREATHE FIRE

The poison that nearly destroyed life
became the fuel that made consciousness possible.

The Poison That Gave Us Thought

In our previous chapter, we witnessed chemistry's greatest shift —the ability to become biology. From the combination of RNA, DNA, and proteins emerged the first true cells—microscopic engines that could grow, reproduce, and evolve. Life had crossed the threshold from mere chemistry into something unprecedented: matter that could remember, learn, and change.

But those first living cells inhabited a world utterly alien to our experience. The skies held no oxygen. The seas were rich with methane and hydrogen sulfide. Lightning crackled through atmospheres that would poison us in seconds. It was a world of chemical possibility, but limited complexity.

Then, in the ancient oceans of a young Earth, tiny blue-green bacteria stumbled upon a discovery that would transform the planet forever. They learned to capture sunlight for energy—and in doing so, they began to exhale a gas that most life found deadly.

This is the story of how life's waste product became its greatest gift, and how the oxygen revolution set the stage for every complex thought you've ever had.

World of the Invisible

For nearly two billion years after life first emerged, Earth belonged to the unseen. Oceans and shorelines were ruled not by creatures with eyes or limbs, but by microbes—bacteria and archaea so small they were invisible to the naked eye. Yet in their hidden dominion, they transformed the planet. Some thrived in boiling hot springs that would melt metal, others flourished in acid pools that could dissolve stone. Still others lived in the crushing depths of the ocean floor, drawing energy from chemical reactions that had powered life since its earliest days.

These microscopic innovators were Earth's first great evolutionary success story. And in many ways, they remain our planet's most successful inhabitants. During their invisible experiments, they discovered nearly every fundamental trick that life would ever learn:

• how to harvest energy from chemicals,

• how to build complex molecules from simple ingredients,

• how to survive in conditions that would destroy any larger organism.

There is something humbling about recognizing that for most of Earth's history, we complex creatures would have been impossible. The energy budget of early Earth was modest—sufficient for microbial success but incapable of supporting the kind of complexity that would eventually give rise to consciousness itself. Life was abundant and diverse, but it was also profoundly limited by the chemistry available to it.

The Blue-Green Revolution

So, around 3.5 billion years ago something extraordinary happened in those ancient seas. A group of bacteria called cyanobacteria discovered something unprecedented: how to capture energy directly from sunlight.

Photosynthesis wasn't entirely new—some bacteria had already learned to use light to power simple chemical reactions. But cyanobacteria took this process to a revolutionary extreme. They learned to split water molecules apart, using the hydrogen to build sugars and releasing the oxygen as waste.

For the first time, organisms could tap into the vast energy of the sun itself. And unlike the limited chemical resources that had powered life before, sunlight was abundant, reliable, and essentially infinite.

Cyanobacteria began to flourish, spreading through Earth's oceans in vast blooms that painted the waters blue-green with life. Along prehistoric coastlines, they built enormous stromatolite 'cities'—towering, layered structures stretching for miles—that stood as monuments to their success. From these living fortresses, they pumped oxygen into the seas and sky at rates never before seen, igniting the planet's greatest atmospheric transformation: the Blue-Green Revolution.

But to most life on the planet, this oxygen was poison.

The Great Poisoning

For over a billion years, Earth's atmosphere had been rich in methane, ammonia, and other gases that we would find toxic today. Most organisms had evolved to thrive in this oxygen-free world, using chemical processes perfectly adapted to their ancient environment.

To these creatures, oxygen was not the breath of life we know today—it was a corrosive poison that disrupted the delicate molecular machinery they depended upon for survival. Imagine if the air you breathed suddenly became acid. That was what cyanobacteria were doing to their world: transforming the very atmosphere that had nurtured life for billions of years.

As cyanobacteria continued to multiply and photosynthesize, they created Earth's first environmental crisis. The Great Oxidation Event, beginning around 2.4 billion years ago, was a slow-motion catastrophe that unfolded over hundreds of millions of years—a timescale that dwarfs human civilization but was rapid by geological standards.

The rising oxygen didn't just threaten individual organisms—it reshaped the entire planet. It reacted with iron dissolved in the oceans, creating vast deposits of rust that settled to the seafloor in distinctive red bands that geologists can still read like pages in Earth's autobiography. It destroyed the methane greenhouse gases that had kept Earth warm, plunging the planet into ice ages so severe that glaciers may have reached the equator.

This was life's first lesson in unintended consequences: how success could become catastrophe, how innovation could threaten the very system that enabled it. It would not be the last time that life would face such a paradox.

The Energy Revolution

For the organisms that could adapt to the new oxygen-rich environment, this gas offered something exceptional: a massive energy bonus.

Burning fuel with oxygen—a process called aerobic respiration —releases far more energy than the older, oxygen-free methods that early life had used. It's the difference between a campfire and a nuclear reactor. Where fermentation might extract a few units of energy from glucose, aerobic respiration could extract over 30 times as much.

This energy windfall enabled something new: large, complex cells that could do sophisticated work. With abundant energy available, cells could afford to build elaborate internal structures, maintain complex chemical processes, and grow to sizes that would have been impossible in the low-energy world of early Earth.

Some pioneering cells made an even more revolutionary discovery They could capture smaller bacteria and put them to work as internal power plants. Instead of trying to develop aerobic respiration themselves, they essentially enslaved oxygen-breathing bacteria and kept them as permanent residents.

This wasn't conquest—it was partnership. The captured bacteria, safe inside their host cells, devoted themselves entirely to energy production. In return, the host cells provided protection and nutrients. Over time, these bacterial partners became so integrated with their hosts that they could no longer survive independently.

We call these captured bacteria mitochondria, and every complex cell on Earth today—including every cell in your body—depends on their descendants for energy. You are powered by the evolutionary descendants of bacteria that learned to breathe oxygen over two billion years ago.

The Birth of Complexity

With their new mitochondrial power plants, cells could afford to experiment with complexity in ways that had never before been possible. They began to develop internal compartments—tiny, specialized chambers where different chemical processes occurred without interfering with each other.

The most important of these compartments was the nucleus—a secure vault where DNA could be stored and protected from the chaotic chemistry of the cell's interior. With their genetic material safely locked away, these new cells could grow larger and more sophisticated without risking damage to their essential instructions.

These revolutionary cells, called eukaryotes (*yoo-CARE-ee-oats*), were the ancestors of every complex organism on Earth today. Plants, animals, fungi, and countless single-celled organisms all descended from those first cells that learned to harness the power of oxygen.

But the partnership between cells and captured bacteria didn't stop with mitochondria. Some cells also captured cyanobacteria and put them to work as internal solar panels. These captured photosynthetic bacteria became chloroplasts—the green machines that power plant cells and ultimately produce the oxygen we breathe today.

Through these ancient acts of cellular cooperation, life learned to turn its greatest crisis into its greatest opportunity. The oxygen that had once threatened to destroy life became the foundation for every complex organism that would ever exist.

The Long Apprenticeship

Even with their new energy sources and complex internal structures, eukaryotic cells remained largely single-celled organisms for hundreds of millions of years. They experimented with different shapes, sizes, and lifestyles. Some became predators, hunting and consuming other cells. Others became decomposers, breaking down dead material and recycling nutrients. Still others remained photosynthetic, building their own food from sunlight.

But perhaps most importantly, some of these cells began to experiment with cooperation on a new scale. Instead of living as individuals, they began to form colonies—groups of cells that worked together while maintaining their individual identities.

These early colonial organisms were like cellular villages, with different individuals taking on specialized roles. Some cells might focus on reproduction, others on feeding, still others on protection from predators. Through this division of labor, colonial organisms could achieve things that no individual cell could accomplish alone.

Over time, some colonies became so integrated that their members could no longer survive independently. The colonial organisms had become multicellular organisms—single beings composed of many cooperating cells.

This transition from single cells to multicellular life was one of the most important innovations in evolutionary history. It opened entirely new possibilities for size, complexity, and sophistication. But it also required something extraordinary: cells that could sacrifice their own reproductive interests for the good of the larger organism.

The Sexual Revolution

Around this time, life made another crucial discovery: sexual reproduction. For billions of years, organisms had multiplied by simple copying, each new individual a near-perfect duplicate of its parent. This method was efficient, but it locked life into sameness. Over time, small accidents in cell division, the fusion of neighboring cells, and the swapping of bits of DNA began to create offspring that were not exact copies. What started as chance exchanges slowly took shape into a new strategy—sexual reproduction.

Sexual reproduction was messier, more complicated, and seemingly less efficient. It required finding a partner, combining genetic material from two different individuals, and producing offspring that were unlike either parent. From a purely practical standpoint, it seemed like a step backward.

But sexual reproduction offered something that asexual reproduction couldn't: genetic diversity. By mixing genes from two parents, sexual reproduction created offspring that were genetically unique. This meant that populations could adapt more quickly to changing conditions, since some individuals would always possess genetic combinations that might prove advantageous.

Sexual reproduction was evolution's way of hedging its bets. Instead of producing many copies of the same genetic blueprint, organisms could produce a variety of different blueprints and let natural selection determine which ones worked best in current conditions.

This innovation would prove crucial for everything that followed. Sexual reproduction accelerated the pace of evolution, enabling life to adapt more quickly to new challenges and opportunities. It set the stage for the explosion of diversity that was about to transform Earth's biosphere.

The Stage is Set

By around 800 million years ago, all the pieces were in place for life's next great transformation. The oxygen revolution had provided abundant energy for complex cellular processes. Eukaryotic cells had developed sophisticated internal structures and the ability to cooperate in multicellular organisms. Sexual reproduction had accelerated the pace of evolutionary innovation.

Earth itself had changed dramatically since the early days of cyanobacteria. The atmosphere now contained significant amounts of oxygen. The climate had stabilized after the ice ages triggered by the oxygen crisis. Ocean chemistry had reached a new equilibrium that could support more complex ecosystems.

Most importantly, predation had emerged as a major evolutionary force. In the early microbial world, organisms had mostly competed for resources through speed and efficiency. But as cells grew larger and more complex, some began to eat others directly. The first predators had appeared, and with them came entirely new selection pressures.

Life was no longer just about finding energy and reproducing efficiently. Now it was also about avoiding being eaten—and about being good enough at eating others to survive. This arms race between predators and prey would drive innovations in mobility, sensing, defense, and intelligence that would transform life on Earth.

Looking Forward

The oxygen revolution had fundamentally altered the trajectory of life on Earth. What had begun as an environmental crisis had become the foundation for complexity, cooperation, and ultimately consciousness itself.

The energy budget that oxygen metabolism provided made possible the large, sophisticated organisms that were about to appear. The cooperation that multicellularity required would become the template for every complex organism that followed. The genetic diversity that sexual reproduction enabled would fuel the adaptive radiations that were about to transform Earth's biosphere.

But perhaps most importantly, the oxygen revolution had taught life its most profound lesson: catastrophe and opportunity are often the same event, viewed from different perspectives. The crisis that nearly destroyed life had become the gift that made complex life possible.

In our next chapter, we will witness the spectacular culmination of these innovations as life explodes into visible complexity for the first time. We will see how the foundations laid during the oxygen revolution enabled the greatest burst of evolutionary creativity in Earth's history—an event so remarkable that scientists call it simply "the Explosion."

7 THE EXPLOSION OF FORM—
WHEN LIFE BECAME VISIBLE

*"Endless forms most beautiful and most wonderful
have been, and are being, evolved."*
— **Charles Darwin**
(Naturalist)

The Invisible Billions

I n our last chapter, we witnessed life's greatest transformation: the oxygen revolution that turned a toxic waste product into the fuel for complexity. We learned how cellular cooperation created the first multicellular organisms, how sexual reproduction accelerated evolutionary innovation, and how predation introduced entirely new selection pressures.

By 600 million years ago, all the pieces were in place for life's next great leap. The energy budget was abundant. The genetic toolkit was sophisticated. The competitive pressures were intense. But life remained largely invisible—soft-bodied creatures that left few traces, living and dying in ancient seas without fanfare or fossil record.

Then something extraordinary happened. In what scientists call the Cambrian Explosion, life suddenly burst into visibility. In just 20 million years—a geological sneeze—nearly every major structural design of the animal body appeared, from shells to jointed legs to segmented worms. It was evolution's first great creative binge, and it changed the rules of existence forever.

The Quiet Before the Storm

For more than three billion years, life on Earth had been a microscopic affair—tiny, unseen, and clinging to existence in oceans, lakes, and damp shorelines. Then, around 600 million years ago, something remarkable stirred beneath the waves. For the first time, large, multicellular organisms appeared, their bodies soft as jelly and their forms unlike anything alive today.

These were the Ediacaran organisms—Earth's first grand experiments in large-scale cooperation between cells. They left no bones or shells to fossilize, only delicate impressions pressed into ancient sea beds. Some spread out as quilted, pillow-like discs that hugged the seafloor, absorbing nutrients directly from the water around them. Others rose like fern fronds, anchored in place but swaying gently in the currents, their leaf-like surfaces maximizing contact with passing food. A few displayed enigmatic, ribbed patterns and branching shapes, as if nature were sketching early blueprints for the body plans that would dominate in the eons to come.

THE RISE OF PREDATION

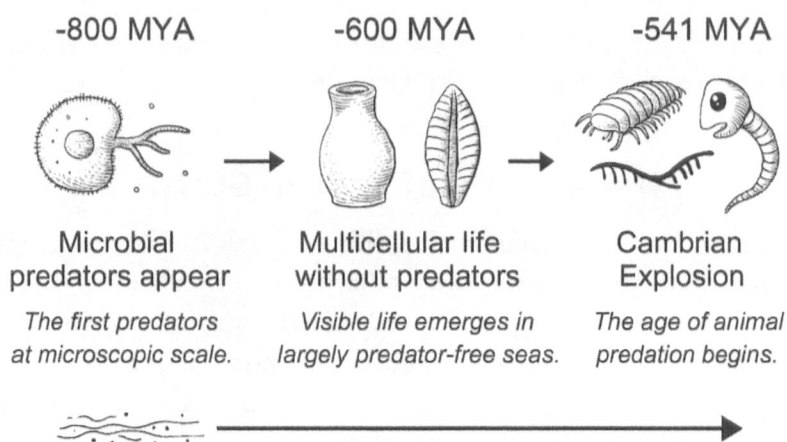

-800 MYA	-600 MYA	-541 MYA
Microbial predators appear	**Multicellular life without predators**	**Cambrian Explosion**
The first predators at microscopic scale.	*Visible life emerges in largely predator-free seas.*	*The age of animal predation begins.*

The rise of predation over deep time: First appearing at the microscopic scale around 800 million years ago, predators initially targeted single-celled life. By 600 million years ago, multicellular organisms flourished in largely predator-free seas. Then, during the Cambrian Explosion (-541 million years ago), animal predators emerged, forever transforming ecosystems and driving evolutionary arms races.

They were silent pioneers, appearing in seas that had only recently grown rich with oxygen. Floating mats of microbial life had dominated Earth for billions of years, but now, around 600 million years ago, strange multicellular organisms began to rise in the shallows. They had no

bones, no shells, and no predators yet to threaten them—only the steady rhythms of tides and sunlight.

Their forms were beautiful in their alien simplicity—leaf-like fronds, quilted discs, and tubular shapes spread across the seafloor like living gardens. But they were profoundly limited; most could not move on their own, had no eyes to sense the world, and no defenses beyond their soft bodies. They lived passively, filtering nutrients from the water or absorbing them directly through their thin skins.

For a brief span in Earth's timeline, they flourished, leaving delicate impressions in the rocks we study today. Yet they would vanish before the Cambrian Explosion, replaced by creatures with shells, movement, and the beginnings of predation. Though their reign was short, it marked a turning point—the dawn of visible life and a preview of the diversity soon to reshape the biosphere.

These quiet creatures were preparing the stage for the greatest creative explosion in the history of life. In their soft bodies, the genetic circuits that would control animal development were being tested and refined. The molecular machinery for building complex body plans was being assembled. The cellular communication systems that would coordinate the behavior of millions of cells were evolving toward their moment of breakthrough.

The Ediacaran period was life's dress rehearsal for consciousness itself. The real performance was about to begin.

Life's First Great Innovation Boom

Around 541 million years ago, something extraordinary happened in Earth's oceans. Life exploded into a riot of form and function that has never been equaled. In just 20 million years—less than half a percent of life's total history—nearly every major group of animals appeared in the fossil record.

Trilobites scuttled across the seafloor on dozens of jointed legs; their compound eyes made of perfectly aligned calcite crystals providing them with sophisticated vision. Anomalocaris, a meter-long predator with grasping arms and a circular mouth full of teeth, cruised through the water column like an alien submarine. Opabinia peered through the murky water with five mushroom-shaped eyes while wielding a bizarre vacuum-cleaner proboscis tipped with a grasping claw.

Hallucigenia walked on seven pairs of spiny stilts, its back bristling with defensive spikes. Wiwaxia resembled a medieval pinecone covered in scales and spines. Pikaia, a small, worm-like creature, carried within its simple body the first known notochord

—the flexible rod that would one day become the backbone of every vertebrate on Earth, including us.

These weren't just variations on existing themes. They were entirely new ways of being alive: animals with shells, spines, claws, and compound eyes. Predators and prey locked in evolutionary arms races. Burrowers that turned the seafloor into Swiss cheese. Filter feeders that strained nutrients from the water with unprecedented efficiency.

The Cambrian seas were a laboratory of biological innovation, testing solutions to problems that life had never faced before. How do you catch prey that can see you coming? How do you defend against predators with crushing claws? How do you process visual information from hundreds of individual lenses? How do you coordinate the movement of dozens of legs?

Evolution answered these questions with a creativity that has never been surpassed.

The Arms Race Begins

The Cambrian Explosion wasn't just about new body plans—it was about new ways of interacting. For the first time in Earth's history, the primary challenge for most organisms wasn't just finding energy or avoiding toxic chemicals. It was dealing with other organisms that wanted to eat them.

Predation drove an evolutionary arms race that accelerated innovation at serious speed. Prey animals developed shells, spines, burrowing behavior, and sophisticated escape responses. Predators evolved crushing claws, drilling mouth parts, improved vision, and hunting strategies.

This wasn't the slow, patient process of adaptation to physical environments that had characterized early evolution. This was rapid, reciprocal evolution where each innovation by one species created immediate pressure for counter-innovations by others.

The trilobites developed some of the most sophisticated visual systems ever created. With compound eyes containing thousands of individual lenses, they could detect movement, judge distances, and even analyze the polarization of light. Some species could see equally well in all directions, making them nearly impossible to ambush.

In response, predators developed new hunting strategies. Anomalocaris evolved flexible front appendages that could grasp struggling prey and powerful jaws that could crush shells. Other predators learned to attack from below, exploiting blind spots in their prey's visual systems.

The seafloor itself became a battlefield. Animals learned to burrow into sediment to escape predators, creating the first three-dimensional ecosystems as life moved not just across surfaces but through them. Others developed the ability to swim rapidly through the water column, turning the entire ocean into hunting grounds.

The Evolution of Seeing

Perhaps the most revolutionary innovation of the Cambrian period was sophisticated vision. While some earlier animals may have been able to detect light and shadow, Cambrian creatures developed the first true eyes capable of forming clear images of their world.

The trilobites were pioneers in this visual revolution. Their compound eyes, like those of modern insects, were made of hundreds or thousands of individual lenses, each contributing a pixel to the overall image. But unlike modern arthropods, trilobite lenses were made of calcite crystal, perfectly aligned to minimize optical distortion.

These crystalline eyes could function underwater with remarkable clarity. Some species had eyes that wrapped around their heads, providing nearly 360-degree vision. Others had eyes mounted on stalks that, like periscopes, could be raised above the seafloor while the rest of the body remained hidden in sediment.

But perhaps most remarkably, some trilobites developed the ability to see underwater with the same clarity that modern animals see in air. Their eyes included sophisticated systems for correcting the optical distortions that water normally creates, allowing them to spot predators and prey with amazing accuracy.

The evolution of vision created entirely new possibilities for behavior. Animals could now hunt by sight, navigate by landmarks, recognize individuals of their own species, and communicate through visual displays. The arms race between predator and prey intensified as seeing creatures faced being seen.

The Burgess Shale Window

Our knowledge of Cambrian life comes largely from a few extraordinary fossil deposits that preserved soft tissues as well as hard parts. The most famous of these deposits is the Burgess Shale in the Canadian Rockies, where a mudslide 508 million years ago buried an entire ecosystem in oxygen-free sediment, preserving it in exquisite detail.

In the Burgess Shale, we see not just shells and skeletons but entire animals: their muscles, guts, gills, and even their last meals. We can study the feeding appendages of predators, the defensive spines of prey, and the delicate branching gills that extracted oxygen from ancient seawater.

These fossils reveal an ecosystem of stunning complexity and alien beauty. Many Burgess Shale animals have no modern relatives—they represent evolutionary experiments that flourished briefly and then vanished, leaving no descendants. Some were as strange as science fiction: *Opabinia*, with five eyes and a long, claw-tipped trunk for grasping prey; *Hallucigenia*, a spiny worm that once baffled scientists

Opabinia, Hallucigenia, and Wiwaxia—three Cambrian creatures that left no modern descendants, showcasing life's early and astonishing experiments in body design.

because it was reconstructed upside down; and *Wiwaxia*, a soft-bodied grazer armored with scales and spikes like a living pincushion. They show us that life once explored possibilities it never returned to—body plans that worked in their time but were eventually abandoned.

Yet within this alien menagerie, we can trace the origins of familiar groups. Early arthropods that would give rise to insects, spiders, and crabs. Primitive chordates that carried the first backbones. Mollusks that would eventually produce snails, clams, and octopuses.

The Burgess Shale and similar deposits are time machines that transport us to life's most creative period, when the basic architecture of animal bodies was being established through trial and error on a planetary scale.

The 1st Mass Extinction– An Ice Age to Kill

The Cambrian Explosion was a brilliant opening act—a burst of innovation that seeded the oceans with strange, new life forms. But over the next 300 million years, evolution continued to sculpt and refine, filling every available niche and inventing new ones.

By the late Ordovician Period, roughly 445 million years ago, Earth's surface looked very different from today. Most of the planet's land was gathered into just two massive continents. To the north lay Laurentia, an ancient landmass that would one day form the core of North America, Greenland, and parts of Scotland. To the south, the colossal Gondwana stretched across vast latitudes.

It was Gondwana's slow, inevitable drift that triggered the crisis. Over millions of years, it crept toward—and eventually settled over —the South Pole. The consequences were dramatic: massive ice sheets began to build, locking away huge amounts of water. Global sea levels plummeted by as much as 70 meters (230 feet), draining the shallow continental seas that had been the cradle of marine life.

The world entered a sudden ice age. Temperatures dropped sharply, ocean currents shifted, and ecosystems collapsed. This

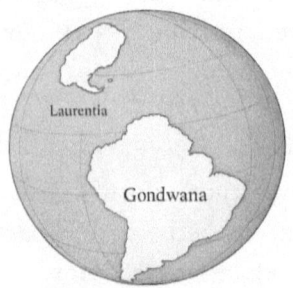

Earth during the Cambrian

Earth (~541–485 million years ago) The supercontinents **Laurentia** and **Gondwana** dominated the globe, separated by vast shallow seas that fostered the explosive diversification of animal life known as the **Cambrian Explosion**.

climatic upheaval—known as the Ordovician–Silurian extinction—was the first great mass extinction in Earth's history.

The toll was staggering—more than 85% of marine species disappeared. Many of the bizarre and experimental body plans born in the Cambrian vanished forever. Even trilobites, once among the most resilient and diverse life forms, suffered massive losses. Early coral reefs, brachiopods, and graptolites—once abundant—were decimated.

And yet, extinction events have always been both an end and a beginning. In the wake of this catastrophe, life rebounded in new and unexpected ways. Early fish evolved jaws, transforming from passive filter-feeders into swift, active predators. The first plants began to anchor themselves to bare rock, breaking the rock down into primitive soils. This green foothold on land created entirely new habitats and altered the planet's atmosphere in ways that would set the stage for future life.

The Ordovician–Silurian extinction was the Great Dying of its time

Early Paleozoic Seafloor Life Shallow seas teemed with diverse communities, including **coral-like reefs**, **brachiopods** (shelled filter-feeders), and **graptolites** (colonial organisms anchored to the seafloor). These ecosystems thrived for millions of years before many lineages were reshaped by mass extinctions.

—a sudden contraction of the biosphere that also cleared the stage for the next great evolutionary act. It was a reminder, written deep in the planet's history, that life's triumphs are never permanent and that even in devastation, opportunity waits for those able to adapt.

Life Moves to Land

The colonization of land was one of evolution's greatest challenges. For over three billion years, life had been confined to aquatic environments where water provided support, protection from radiation, and a medium for reproduction. Moving onto land required solutions to entirely new problems.

Plants led the way, developing specialized structures to extract water from soil, transport it through their bodies, and prevent its loss to the dry air. They created waxy coatings to protect against desiccation and rigid cell walls to support their weight against gravity.

Early land plants were simple, low-growing organisms that stayed close to water sources. But they began to transform terrestrial environments in profound ways. Their roots broke down rock, creating the first true soils. Their photosynthesis pumped oxygen into the atmosphere while removing carbon dioxide. Their presence created shade, shelter, and new habitats for other organisms.

Animals followed plants onto land, initially as temporary visitors that returned to water to reproduce. Early arthropods—ancestors of insects and spiders—developed waterproof exoskeletons and more efficient respiratory systems. The first vertebrates to venture onto land were lobe-finned fish that could breathe air and use their muscular fins to move across mudflats and through shallow water.

These early terrestrial animals faced challenges that their aquatic ancestors had never encountered. How do you support your body weight without the buoyancy of water? How do you breathe air instead of extracting oxygen from water? How do you reproduce without an aquatic environment for your eggs and larvae?

Evolution solved these challenges through gradual modification of existing structures and the invention of new ones. Over millions of years, certain lobe-finned fishes strengthened the bones at the base of their paired fins until they could bear weight; those bones became the humerus, radius, and ulna of early tetrapods and eventually of every walking animal. The swim bladder—first a buoyancy organ of fish—subtly changed its blood supply and tissue, giving rise to primitive lungs so that air-breathing fish such as lungfish and the ancestors of amphibians could survive in oxygen-poor waters. Among the first true land dwellers were the amniotes—early reptiles whose embryos developed inside a fluid-filled sac called an amnion. They evolved the amniotic egg, with a protective shell and specialized membranes that conserved water and nourished the embryo, and from this foundation came another leap: the evolution of live birth in many reptiles and mammals, freeing reproduction entirely from dependence on standing water.

The 2nd Mass Extinction: Oceans in Collapse

Around 375 million years ago, life faced its second great extinction crisis. The Devonian extinction was different from the Ordovician event. Instead of a sudden ice age, it involved a more gradual but equally devastating change in ocean chemistry.

As plants continued to colonize the land, they accelerated the weathering of rocks and increased the flow of nutrients into rivers and oceans. These nutrients fueled massive algal blooms that consumed oxygen as they decomposed, creating vast dead zones in ancient seas.

The extinction played out over millions of years, with repeated pulses of species loss as ocean ecosystems collapsed and recovered multiple times. Many groups of marine animals that had dominated Paleozoic seas disappeared forever, including several major groups of fish and marine invertebrates.

But once again, extinction paved the way for innovation. The groups that survived the Devonian crisis diversified rapidly in its aftermath. Amphibians—vertebrates that could live both on land and in water—began their radiation. Insects took to the air for the first time. Forests of seed-bearing plants spread across the continents.

The 3rd Mass Extinction
—The Great Dying

The third mass extinction—and the most devastating in Earth's history—struck at the end of the Permian Period, about 252 million years ago. Known as the Great Dying, it wiped out more than 90 percent of marine species and about 70 percent of land-dwelling vertebrates.

Its trigger was a deadly partnership between massive volcanic eruptions in what is now Siberia and the slow assembly of the supercontinent Pangaea. The Siberian eruptions poured vast amounts of carbon dioxide and other gases into the air, driving global temperatures sharply upward. Warmer oceans held less dissolved oxygen to begin with, and the altered climate increased rainfall and rock weathering on land, sending a flood of nutrients into the sea. Those nutrients fueled huge blooms of plankton and algae.

PANGAEA

Around 250 million years ago, nearly all of Earth's landmasses were joined into a single giant supercontinent called **Pangaea.** This vast landform stretched from pole to pole and dramatically shaped climate, ocean circulation, and the evolution of life during the late Paleozoic and early Mesozoic eras.

When these blooms died and sank, bacteria feasted on the organic remains, consuming oxygen far faster than it could be replenished. At the same time, the newly formed Pangaea reduced shallow coastal seas and slowed ocean circulation, further limiting the mixing of oxygen-rich surface water with deeper layers. The deep ocean became anoxic—devoid of oxygen.

In these dark, stagnant waters, microbes that thrive without oxygen took over. They used sulfur instead of oxygen to break down organic matter, releasing hydrogen sulfide, a poisonous gas. As oxygen-free zones spread upward, hydrogen sulfide bubbled into surface waters and sometimes into the air, tainting the atmosphere and damaging the ozone layer.

The result was catastrophic: ocean temperatures climbed by as much as 10°C, waters turned acidic, and vast stretches of sea became toxic and suffocating. Together, warming, acidification, and oxygen depletion created the perfect storm that caused the greatest biological collapse our planet has ever endured.

Life came closer to complete extinction than at any other time in Earth's history. Ecosystems collapsed across the planet. Food chains that had existed for hundreds of millions of years were severed. The diversity that had been building since the Cambrian Explosion was nearly wiped clean.

Yet even this ultimate catastrophe became a doorway to new possibilities. In the devastated world that emerged from the Great Dying, a small group of reptiles called archosaurs began to diversify. They would give rise to crocodiles, pterosaurs, and most importantly, dinosaurs.

The Age of Dragons Begins

From the ruins of the Great Dying, a new lineage began to rise—small, agile reptiles called *archosaurs*. They were not the largest creatures of their time, nor the most fearsome, but they carried within them a suite of adaptations that would prove world-changing: an upright posture for efficient movement, a metabolism that hinted at warm-blooded activity, and lungs built for a one-way flow of air. Unlike the back-and-forth breathing of mammals, where each breath mixes fresh and stale air, archosaur lungs worked like a steady wind tunnel. Fresh air entered at one end and exited at another, keeping oxygen moving across lung surfaces all the time. This constant stream let archosaurs draw far more oxygen from every breath, powering active lives even when the atmosphere held less oxygen than today. That extraordinary design has largely vanished, surviving only in their distant descendants—birds and crocodilians.

Over millions of years, some archosaurs gave rise to crocodilians, others to winged pterosaurs, and one branch to the most celebrated rulers in Earth's history—the dinosaurs. At first, they were modest herbivores and hunters in a world still recovering from catastrophe. Yet their efficiency, adaptability, and evolutionary potential made them ready for opportunity when it came.

That opportunity would arrive with another planetary crisis—the fourth great mass extinction. It would sweep away many of their competitors and open the door to a reign that would last for more than 160 million years.

The 4th Extinction and the Rise of Dinosaurs

The world that rose from the Great Dying was a harsh and unfamiliar place, yet life proved resilient. Over the next 50 million years, ecosystems slowly rebuilt themselves. The supercontinent Pangaea sprawled from pole to pole, creating vast arid interiors and seasonal extremes. Early dinosaurs shared the land with many other reptilian lineages—armored herbivores, crocodile-like predators, and strange, long-necked grazers.

In the seas, marine reptiles like ichthyosaurs and plesiosaurs hunted in warm, shallow waters, while ammonites and new species of coral reefs flourished.

But stability in Earth's history is always temporary. Around 201 million years ago, the planet convulsed again. The breakup of Pangaea began with one of the largest volcanic events in Earth's history—the eruption of the Central Atlantic Magmatic Province. For hundreds of thousands of years, immense lava flows covered millions of square kilometers, releasing staggering amounts of carbon dioxide and sulfur gases into the atmosphere.

The consequences were global and swift. Greenhouse warming

pushed temperatures to extremes, while ocean acidification dissolved the very shells of marine creatures. Vast regions of the sea became starved of oxygen. On land, sudden climate swings brought drought, flooding, and habitat collapse. In the turmoil, 70-80% of all species vanished. Entire groups of large amphibians and many competing archosaurs disappeared, along with numerous marine reptiles and invertebrates.

Triassic shoreline and sea life. **Top (land, left → right):** aetosaur-like armored herbivore; early sauropodomorph long-necked grazer; crocodile-like predator (phytosaur). **Below (shallow sea, left → right):** ichthyosaur; plesiosaur; nothosaur-like marine reptile; ammonite, with early corals on the seafloor.

This extinction reshaped the balance of power. With their rivals gone, the dinosaurs—still relatively small and modest in form—found themselves in a world full of vacant ecological roles. Over the next ten million years, they would diversify into towering sauropods, horned giants, plated herbivores, and swift, intelligent predators. The stage was set for the true Age of Dinosaurs, which would dominate the Jurassic and Cretaceous for over 160 million years.

A Reflection on Creativity and Crisis

The story of the Cambrian Explosion and the great extinctions that followed reveals one of evolution's deepest truths: creativity and crisis are intimately connected. Life's greatest innovations have often emerged not during stable periods, but in the aftermath of catastrophic challenges.

The Cambrian Explosion followed the environmental upheaval of the oxygen revolution. The colonization of land was spurred by changing sea levels and climate. The rise of dinosaurs emerged from the devastation of the Great Dying. Each time, what seemed like an ending became a beginning.

This pattern suggests something profound about the nature of innovation itself—that genuine creativity often requires the destruction of existing structures and the pressure of genuine need. Life's most beautiful solutions have been responses to its most desperate problems.

In our own time, as we face environmental challenges that threaten to reshape civilization, we might find hope in this ancient pattern. Crisis has always been evolution's editor and creativity's catalyst. The same pressures that eliminate some possibilities often create space for entirely new ones to emerge.

We are the current inheritors of this long conversation between challenge and innovation, crisis and creativity. Our response to the difficulties we face today will determine what new chapters get written in the ongoing story of consciousness in the cosmos. The universe is still experimenting with what awareness can become, and we are both its latest experiment and its most conscious participants in that experiment.

8 WHEN GIANT REPTILES RULED

Dominance and Shadow

In our last chapter, we left the world forever changed by the Triassic–Jurassic extinction. Most of the dinosaurs' competitors were gone, and the survivors now found themselves in a world rich with opportunity. It was the dawn of their true ascendancy.

For more than 160 million years, dinosaurs would dominate the land with a diversity and splendor unmatched in Earth's history. From towering plant-eating sauropods that could strip a forest in days, to sleek, sharp-eyed predators that hunted with speed and cunning, they reshaped entire ecosystems simply by existing.

But while the dinosaurs' age is remembered for its scale and spectacle, another story was unfolding in the quiet spaces they left untouched. In the shadows and undergrowth, small, warm-blooded mammals were refining the traits—adaptability, cooperation, and care—that would one day outlast the giants themselves.

The Early Jurassic World

By the opening of the Jurassic Period, around 200 million years ago, the Earth had settled into a warmer, wetter climate. The great deserts of the Triassic began to shrink, giving way to lush forests of conifers, cycads, and towering tree ferns. Shallow inland seas and broad river valleys created a mosaic of fertile habitats across the continents. In this revitalized world, dinosaurs began their first great expansion. Freed from many of their former competitors, they radiated into every available niche.

Alongside them, flying reptiles ruled the skies. Pterosaurs ranging from sparrow-sized insect hunters to wide-winged fish catchers went gliding over lakes and coasts. In the seas, marine reptiles like ichthyosaurs and plesiosaurs thrived, their streamlined bodies built for speed and precision in pursuit of prey.

It was a time of ecological balance—predators and prey locked in dynamic tension, plants and herbivores shaping each other's evolution, and every corner of the landscape teeming with life. The Jurassic had opened as an age of opportunity, and dinosaurs were making the most of it.

The Dinosaur Innovation

With their rivals gone, dinosaurs surged into an age of rapid diversification. They evolved into an astonishing range of forms— Some grew into towering herbivores that could browse treetops no other animal could reach, while others became swift, agile predators that prowled the forest margins and open floodplains. Their anatomical innovations, from highly efficient lungs to lightweight but powerful skeletons, allowed them to thrive in environments as varied as dense forests, open plains, and coastal wetlands. This adaptability would define their reign for the next 160 million years.

Many dinosaurs were probably warm-blooded, able to maintain constant body temperatures that supported high levels of activity. This gave them significant advantages over their cold-blooded competitors, especially in variable climates or during seasonal changes.

But perhaps most importantly, dinosaurs developed remarkable diversity in size, shape, and lifestyle. While early forms were relatively small, the group quickly began to explore the full range of ecological possibilities available to terrestrial animals.

The Giants of the Jurassic

In this fertile world, some dinosaurs began to experiment with gigantism on a scale never before seen on land. The sauropods—long-necked herbivores like Brachiosaurus, Diplodocus, and Apatosaurus—grew to extraordinary sizes, with some species reaching lengths of over 30 meters and weights approaching 80 tons.

These giants were possible because dinosaurs had solved the engineering challenges of extreme size. Their bones were hollow, which drastically reduced weight yet maintained strength. Their long necks allowed them to browse vegetation across wide areas without moving their massive bodies. And their sophisticated digestive systems could extract nutrients from low-quality plant material.

The sauropods weren't just large—they were intelligent about being large. They likely lived in herds that provided protection from predators and shared knowledge about food sources. They may have migrated seasonally, following the growth of vegetation across vast distances.

The Predators' Response

The evolution of giant herbivores created opportunities for equally formidable predators. Theropod dinosaurs—the bipedal carnivores that included everything from chicken-sized Compsognathus to the massive Tyrannosaurus rex—developed increasingly sophisticated hunting strategies and more powerful killing tools.

Allosaurus vs. Giant Sauropod — Jurassic Predator and Prey An Allosaurus launches a powerful attack on a massive sauropod, using clawed hands and serrated teeth while smaller theropods like Compsognathus watch from the forest edge.

Allosaurus, the apex predator of the Jurassic, combined size, speed, and intelligence in ways that no land predator had achieved before. With powerful jaws filled with serrated teeth, grasping hands armed with large claws, and legs built for pursuit, it could bring down prey much larger than itself.

These hunters also developed keen senses, including excellent vision and possibly acute hearing. Some species may have hunted in coordinated packs, using communication and teamwork to overwhelm large prey.

As the Jurassic gave way to the Cretaceous, theropods continued to innovate. Many became specialized for specific types of prey—fish-eaters with crocodile-like skulls, egg-thieves with long fingers and sharp claws, giant killers with bone-crushing jaws.

Perhaps most remarkably, some theropods began to develop feathers. Initially used for display or temperature regulation, feathers would eventually enable the evolution of powered flight, creating an entirely new category of dinosaur.

The Flowering Revolution

The Cretaceous period, beginning 145 million years ago, brought another revolutionary change to terrestrial ecosystems: the evolution and rapid spread of flowering plants.

Angiosperms—plants that produce flowers and enclosed seeds —had several advantages over the conifers and ferns that had dominated earlier forests. These new plants could reproduce more quickly, adapt more rapidly to changing conditions, and form more diverse ecological relationships with other organisms.

The co-evolution of flowering plants and insects created entirely new forms of complexity. Plants developed elaborate flowers to attract pollinators, while insects evolved specialized mouthparts and behaviors to exploit floral resources. This partnership accelerated the diversification of both groups and increased the overall productivity of terrestrial ecosystems.

Dinosaurs adapted to these changing plant communities with remarkable creativity. Duck-billed hadrosaurs developed sophisticated groupings of stacked teeth called dental batteries that could process tough plant material. Horned ceratopsians like Triceratops evolved massive skulls and powerful jaw muscles to deal with fibrous vegetation. Long-necked sauropods continued to exploit high-canopy resources that other herbivores couldn't reach.

The Cretaceous became the peak of dinosaur diversity. Over 700 species are known from this period, representing an extraordinary range of body sizes, feeding strategies, and ecological roles. From tiny, bird-like predators to armored tanks the size of buses, dinosaurs had filled virtually every terrestrial niche available to large animals.

Life in the Shadows

While dinosaurs dominated the visible landscape of the Mesozoic, a very different group of animals was quietly establishing itself in the shadows: the mammals.

The first mammals appeared during the Late Triassic, around the same time as the earliest dinosaurs. But unlike their reptilian contemporaries, early mammals remained small—most were the size of mice or rats, and even the largest were no bigger than modern house cats.

This small size wasn't a limitation—it was an evolutionary advantage in a world dominated by dinosaurs. Small mammals could exploit niches that were unavailable to larger animals. They could live in burrows that provided protection from predators and temperature extremes. They could feed on insects, seeds, and other small food items that larger animals couldn't efficiently harvest.

Most importantly, mammals developed a suite of characteristics that would prove incredibly valuable when the world changed again. They were warm-blooded, able to remain active during cool nights when cold-blooded predators became sluggish. And they had sophisticated sense organs, including acute hearing and smell that allowed them to navigate in darkness.

Perhaps most significantly, mammals evolved complex parental care behaviors. Unlike reptiles, that typically abandoned their eggs after laying, mammalian mothers nour-ished their young with milk and provided extended care that dramat-ically increased the survival rates of their offspring.

Early Mesozoic mammals: a mother nurses her young in a moonlit burrow while others climb and forage beneath conifers and ferns, surviving in niches that giant dinosaurs

The Mammalian Advantage

What made mammals special wasn't just their physiology—it was their minds. Living in a world dominated by much larger predators, early mammals had to out-think danger. They evolved sharper senses and quicker learning: keen hearing to detect the faint rustle of a hunting dinosaur, an acute sense of smell to track food in the dark, and a larger, more layered brain that stored memories and linked experiences. They could remember safe routes to hidden burrows, learn from a single encounter with a predator, and adjust foraging times according to shifting conditions. These flexible behaviors—far beyond simple reflex—gave mammals a survival edge that pure size or strength could never match.

Mammalian brains were larger relative to their body size than reptilian brains, with expanded areas in the brain devoted to sensory processing and behavioral control. This neural sophistication allowed mammals to adapt quickly to changing conditions and exploit new opportunities as they arose.

Many mammals also began to develop social behaviors that increased their survival chances. Some species lived in family groups where adults shared information about food sources and dangers. Others formed larger communities that provided collective protection against predators.

These social innovations required new forms of communication and cooperation. Mammals developed complex vocal repertoires, scent-marking systems, and body language that allowed them to coordinate group activities and maintain social bonds.

The Flying Dragons

Perhaps the most spectacular dinosaur innovation was the evolution of powered flight. Starting with small, feathered theropods in the Jurassic, some dinosaur lineages began to experiment with gliding and eventually powered flight.

The evolution of avian flight was a gradual process that required multiple innovations working together. Feathers provided insulation and later became flight surfaces. Hollow bones reduced weight while maintaining strength. Powerful flight muscles developed to provide the energy needed for sustained flight.

The first birds, like Archaeopteryx, were essentially flying dinosaurs—retaining many reptilian features like teeth, long bony tails, and clawed wings. But they opened entirely new ecological possibilities, allowing dinosaurs to exploit aerial environments that had previously been dominated by pterosaurs.

As birds diversified through the Cretaceous[PM2] , they developed many of the characteristics we associate with modern avians: advanced flight capabilities, sophisticated navigation systems, complex social behaviors, and elaborate mating displays.

The evolution of flight also required enhanced intelligence. Flying animals must process complex three-dimensional information, make rapid decisions about landing sites and obstacles, and often coordinate their movements with other members of their species.

Signs of Intelligence

By the end of the Cretaceous, both dinosaurs and mammals were showing signs of increasing intelligence. Some dinosaur species built elaborate nests and engaged in complex parental care behaviors. Others may have used tools or developed sophisticated communication systems.

Archaeopteryx—an early Jurassic bird retaining dinosaur traits such as teeth, clawed wings, and a long bony tail, displaying the feathered innovations that marked the dawn of powered flight.

Troodon, a small theropod with large eyes and a relatively large brain, showed many characteristics that suggest high intelligence. Its brain-to-body ratio was similar to that of modern birds, and it had highly developed areas associated with sensory processing and motor control.

Some duck-billed dinosaurs may have engaged in complex vocal communication, using their elaborate nasal passages and crests to produce a wide range of sounds. Fossilized trackways suggest that some species traveled in organized herds with specific social structures.

Meanwhile, mammals were developing even more sophisticated cognitive abilities. Some species showed evidence of problem-solving skills, social learning, and flexible behavioral responses to novel situations.

The 5th Extinction— The Cosmic Accident

For 160 million years, dinosaurs ruled the Earth with remarkable success. They had survived multiple lesser extinction events, adapted to changing climates, and diversified into hundreds of species. They seemed perfectly positioned to continue their dominance indefinitely.

Then, 66 million years ago, a mountain-sized asteroid slammed into the Earth near what is now the Yucatan Peninsula in Mexico.

The impact released energy equivalent to billions of nuclear weapons, instantly vaporizing the asteroid and excavating a crater over 150 kilometers wide. Molten rock was blasted into the atmosphere, raining down across continents and igniting global wildfires. Dust and sulfur compounds blocked sunlight for months, plunging the planet into a cold, dark winter.

During this impact winter, photosynthesis ceased. Plants died. Then herbivores starved. And the carnivores followed. Within a geologically brief period, 75% of all species on Earth went extinct, including every non-avian dinosaur.

The age of giants was over.

The Inheritance

In the devastated world that emerged from the impact winter, small size became a survival advantage. The mammals that had lived in dinosaurs' shadows suddenly found themselves in a world with vastly reduced competition and abundant ecological opportunities.

Mammalian characteristics that had been merely useful in the Mesozoic became essential in the post-impact world. Warm-bloodedness allowed survival in the cooler post-impact climate. Small size meant lower energy requirements during that period of reduced food availability. Also, behavioral flexibility allowed rapid adaptation to changed conditions.

Most importantly, mammals possessed something that most dinosaurs had lacked: sophisticated social bonds and parental care systems that could help communities survive through difficult periods.

As the world slowly recovered from the impact catastrophe, mammals began to diversify explosively. Within a few million years, they had evolved forms ranging from tiny insectivores to large browsers, from aquatic specialists to gliding forms that presaged the evolution of bats.

Early Cenozoic recovery: small, warm-blooded mammals forage, nest, and care for their young in a cooler, ash-darkened landscape, seizing new ecological opportunities after the dinosaurs' extinction.

But perhaps most significantly, some mammalian lineages began to develop larger brains and more complex social behaviors. The foundations were being laid for the evolution of intelligence, consciousness, and ultimately, the capacity to understand the very history we've been tracing.

Looking Forward

The extinction of the dinosaurs marked the end of one chapter in life's story and the beginning of another. The age of giants gave way to the age of minds, as mammals began to explore possibilities that size and strength alone could never achieve.

In our next chapter, we will follow the mammalian radiation that followed the asteroid impact, watching as different lineages experimented with various approaches to survival in the post-dinosaur world. We will see how some chose the safety of herds, others the cooperation of packs, and still others the complex alliances that would eventually give rise to consciousness itself.

The giants had ruled through dominance. Their successors would rule through understanding.

A Reflection on Dominance
and Adaptability

The story of dinosaur success and extinction offers profound lessons about the nature of dominance and the importance of adaptability in an ever-changing universe. For over 160 million years, dinosaurs were evolution's greatest success story—diverse, abundant, and seemingly invincible masters of their world.

Yet their very success may have contributed to their vulnerability. Specialized for life in stable, warm climates, many dinosaur lineages were poorly equipped to handle the rapid environmental changes triggered by cosmic accident. Their large size, which had been an advantage in competition with other species, became a liability when resources became scarce and the world turned cold and dark.

In contrast, the small mammals that inherited the post-impact world succeeded not through dominance but through adaptability. Their behavioral flexibility, social cooperation, and efficient physiology allowed them to survive conditions that eliminated their larger, more specialized competitors.

This pattern echoes through cosmic history: the characteristics that make a system successful in stable conditions may be different from those that ensure survival through crisis. In ecosystems and civilizations, adaptability and cooperation often prove more valuable than size, strength, or rigid specialization when the universe presents unexpected challenges.

As we face our own period of rapid change, we might remember the lessons written in deep time: that survival belongs not to the mightiest or most rigid, but to those who can adapt, cooperate, and think their way through unprecedented challenges. The universe doesn't favor the dominant—it favors the wise.

9 THE RISE OF THE GROUP —HOW BELONGING BECAME SURVIVAL

"No man is an island, entire of itself."
— John Donne
(Poet)

After the Fire

A round 66 million years ago, a six-mile-wide rock from space was hurtling toward Earth at 27,000 miles per hour—as fast as the International Space Station orbits Earth. When it struck the Yucatán Peninsula, the impact unleashed more energy than all the world's volcanoes, earthquakes, and bombs combined. Oceans boiled at the coasts. Shockwaves raced around the globe. Fires consumed forests. The sky dimmed beneath a shroud of ash, and for months, perhaps years, sunlight failed to reach the ground. Photosynthesis stopped. Food chains collapsed.

Within a geological heartbeat, 75% of all life was gone— including the dinosaurs, rulers of the land for over 160 million years.

In the quiet that followed, small survivors emerged. The mammals. Warm-blooded and sharp-sensed, they endured the darkness by scavenging, burrowing, and caring for their young. When the skies cleared, they stepped into a world reset—one where the old giants were missing, and evolution could write a new chapter. From that moment, the long arc of consciousness began again.

Solitary Survival in a Fragile World

The Paleocene epoch, 66–56 million years ago, was Earth's recovery room. Forests began to regrow across scarred continents. Rain returned. Rivers flowed again. Volcanoes, once destructive, now released carbon dioxide into the air, warming the planet back toward pre-impact levels. Slowly, sunlight pierced the haze, and life emerged from the shadows.

But the biosphere was fragile, delicate and healing. In this uncertain world, mammals multiplied into a variety of forms: rodent-like foragers sniffing in the leaf litter, gliding insect-eaters soaring from branch to branch, hoofed browsers picking their way through the undergrowth. Yet despite their diversity, most followed the same ancient rule: stay alone.

Solitude offered clear advantages. No need to share food. Less risk of disease. Fewer fights over territory or mates. And for millions of years, it worked. Even today, 85-90% of all animals live solitary lives. Among mammals, roughly two-thirds prefer to be alone. In the animal kingdom, solitude is not loneliness—it is strategy, refined through countless trials.

Still, evolution is never static. From time to time, a new path emerges—not because nature seeks improvement, but because change is inevitable. And sometimes, these new paths lead to entirely different forms of survival.

As the Paleocene gave way to a new epoch, the seeds of such a shift were stirring. In the hidden corners of the forests—between roots and vines, shadows and canopy—a new way of life was preparing to take shape. Not just survival of the body, but the awakening of the mind.

The Greenhouse Earth

By the dawn of the Eocene Epoch, around 56 million years ago, Earth had both recovered from the cataclysm that ended the dinosaurs and blossomed into a planet both alien and extraordinary. Global temperatures soared, reaching 10-12°C (18-22°F) warmer than today. The poles, once frozen, were free of permanent ice. Sea levels rose by hundreds of feet, flooding coastal plains and pushing warm, shallow seas into the heart of continents.

Massive volcanic eruptions, combined with the sudden release of carbon from ocean sediments and melting permafrost, pumped vast amounts of methane and carbon dioxide into the atmosphere. CO_2 levels were likely two to five times higher than today, fueling one of the warmest and most prolonged climate events in Earth's recent history. The world had become a greenhouse.

Across nearly every latitude, lush rainforests unfurled like green banners. From equatorial Africa to what is now northern Canada, the Earth was cloaked in a steaming tapestry of life. Mangroves thrived in Wyoming. Crocodiles basked in Arctic rivers. Ferns, cycads, and flowering plants stretched toward the hazy sky in endless green.

And in the shadows of this tangled canopy, something remarkable was taking shape—a new kind of mammal was rising.

These mammals were descendants of tiny, nocturnal insect-eaters—shrew-like creatures that had survived the dinosaurs by living in secrecy and silence. Their tools were their senses: large eyes for seeing in the dark, flexible fingers for grasping, and a diet rich in insects and fruit. Over tens of millions of years, natural selection honed these traits with precision. The result was the first true primates.

Small—about the size of squirrels—but built for life in the trees, they had forward-facing eyes for depth perception, opposable thumbs for grasping, and long tails for balance as they bounded from branch to branch. Their world was not flat, but vertical. Above them, birds of prey circled. Below, snakes slithered silently.

A pint-sized pioneer of the primate lineage clings to a branch in the warm forests of the Eocene epoch, some 55 million years ago. No bigger than a squirrel, this tiny tree-dweller represents one of evolution's most consequential experiments —the early ancestors that would eventually give rise to all modern primates, from lemurs to humans.

Around them, moss-slick branches tested every leap.

To survive here required more than reflex. It demanded anticipation. Memory. Trust. This was a world that rewarded awareness—not just of the environment, but of others. And so, in this steaming greenhouse Earth, a quiet revolution was beginning. Sensation was becoming cognition. In the high, trembling branches of ancient forests, the first flickers of social intelligence sparked into being.

The First Sparks of Feeling

These early primates developed more than nimble fingers and agile limbs. As they climbed, leapt, and navigated the green cathedral of the Eocene canopy, something deeper was beginning to stir within them. They weren't just reacting to the world—they were beginning to feel it, and to feel each other.

Mothers didn't simply give birth and move on. They carried their infants for months, even years—cradling, protecting, and soothing them. This bond was more than instinct; it was devotion. And it wasn't limited to mothers. When an infant cried, others noticed. Some responded. A scream in the treetops wasn't just a noise—it was a call that triggered what we might now call concern. These were the earliest echoes of empathy.

One of the most profound behaviors to emerge was grooming. At first glance, it may seem like simple hygiene, but it was far more than that. To groom another was to offer peace; to accept grooming was to place trust. Fingers threading through fur calmed nerves, settled disputes, and bound individuals together in invisible cords of loyalty.

Even a glance began to carry meaning. A look could warn, welcome, threaten, or reassure. In that flash of eye contact, something passed between minds—not words, but understanding. And when danger crept close, the forest lit up with alarm calls. High-pitched, sharp, and distinct, they raced from tree to tree like electricity. Others didn't need to see the threat themselves—they reacted because they trusted the signal. One mind sparked another, and a call became a shared decision. A group survived.

These creatures weren't self-aware in the way modern humans are. They didn't imagine the distant future or reflect deeply on the past. But something had changed. They were aware. They felt heat and cold, hunger and safety. They recognized friend and foe. This was the first layer of consciousness: sensory awareness—the ability to feel the world and respond to it.

It wasn't thinking in the abstract. It was experiencing. Yet this simple awareness marked a turning point, allowing individuals to respond not only to their environment but to each other. A rustle meant flee. A touch meant stay. A cry meant danger. Those who could read these signals—who could feel what others felt—had the advantage. They avoided harm, formed alliances, and survived. Those who couldn't often fell behind, missing the signs that spelled life or death.

In these quiet forests, emotional intelligence was taking root. It wasn't yet language. It wasn't yet memory. But it was the ember of connection, the early warmth of mind.

The Pre-language of the Forest

High in the swaying canopy, where wind whispered through leaves and shadows danced across bark, this new form of communication continued to take shape. Not language as we know it, but something older—a transmission of information woven from glances, grunts, and gestures.

In these tight-knit primate bands, meaning lived in motion. A sharp, unblinking flash of the eyes could halt a young male in his tracks without a single sound. The message was clear: don't challenge me. Dominance was enforced not just through force, but through understanding—and understanding didn't require words.

Sounds filled the air: soft coos of contentment, harsh barks of irritation. These weren't random noises; they carried mood and hinted at intention. The tone of a grunt could signal whether to approach...or stay away. Over time, the signals became more refined. A snake's slither prompted one kind of alarm. An eagle's shadow prompted another. Each sound sparked the right reaction: freeze, flee, or climb.

Evolution began to fine-tune vocal cords, ears, and brains— favoring those who could make the right sound and those who could decode it quickly. But not every sound was about danger. Many were about connection. Grooming, once purely practical, became a tool of diplomacy. A touch at the right moment could soothe jealousy; a grooming session could repair a fractured bond. To groom another was to say: I choose you. You matter. You are safe with me.

Meaning flickered in every call, every touch, every glance. This was not yet language, but it was communication—a pre-linguistic vocabulary of rhythm, tone, and expression. Those who could interpret these signals, who could read the emotions behind them, held a powerful advantage. The seeds of language were being planted in the treetops, long before the first story was told.

Why Groups Began to Matter

Most mammals, even in the lush forests of the Eocene, still preferred solitude. Nocturnal foragers, burrowing creatures, and territorial hunters lived and hunted alone, as their ancestors had for millions of years. Even among primates, some species remained isolated or drifted through loose, shifting bands that broke apart with the seasons.

But in the rainforest canopy, a quiet transformation had begun. The treetops were crowded and unpredictable, full of perils from above and below. Fruit trees bloomed only briefly, and infants needed months—sometimes years—of care before they could fend for themselves.

In this demanding world, a new strategy emerged: group living. These were intentional communities shaped by memory, trust, and shared survival. More eyes watched for danger. More hands cared for the young. More minds remembered where the ripest fruit would return after the rains.

To live in a group meant someone could warn you, teach you, help you. But it also meant risk. You had to know who could be trusted, and sense who might turn aggressive. You had to forgive, groom, follow—and sometimes lead. This was more than herding. It was relational survival.

Seasonal shifts, scattered food, and rising stakes drove this change. Over millions of years, evolution began to rewire the mind —not for dominance alone, but for belonging. Those who bonded lived longer. Those who connected passed on their genes. Those who formed alliances, shared vigilance, and built trust, thrived. Being part of a group was no longer optional; it was as essential as breath.

A Slow Rewiring of the Brain

Over countless generations, the pressures of social life reached inward, shaping not just behavior but the very structure of the brain. To survive in a group, memory had to expand. It wasn't enough to remember where the fruit ripened or the leopard prowled—you had to recall who helped you last week, who stole your food, who groomed you, and who held your infant when you were too tired to climb.

The brain began building a new kind of map—a mental record of relationships. Who was loyal. Who was jealous. Who could be trusted. The emotional centers deep in the brain grew more intricate. Fear and hunger were joined by affection, anxiety, anticipation, and resentment. These weren't abstract ideas but physical experiences, floods of chemicals through blood and bone.

Oxytocin surged when two individuals touched or groomed, reinforcing trust and safety. Dopamine spiked during positive social encounters, encouraging kindness and connection. Emotion became not just instinct but preference. Early primates began to choose—seeking favored companions, remembering betrayals, and mourning losses.

With this came a second layer of consciousness: internal awareness—the ability to feel not only the world outside, but the one within. Love, jealousy, joy, sadness—these feelings were real. And they guided action, shaped memory, and altered the course of survival. Biology was becoming mind.

When the World Began to Chill

Around 34 million years ago, the Earth's climate cooled by 4-5°C (7-9°F). It was enough to redraw ecosystems. The steaming rainforests that had blanketed much of the world retreated, replaced by open woodlands, rolling grasslands, and scattered patches of forest hemmed in by expanding plains.

For the first time in over 100 million years, ice returned—forming in the highlands of Antarctica and creeping outward into the seas. Seasonality became the new rhythm: dry months, lean winters, wet seasons, drought.

Fruit was no longer abundant year-round. It came in bursts, here one month, gone the next. Rain came late. Trees bore nothing. Many primates couldn't adapt and vanished from the fossil record. But others adapted, forming smaller, tighter, more cooperative groups. They shared food, took turns watching for danger, and raised young together.

The group became more than protection. It became a collective mind, a living memory, a survival machine. With cooperation came complexity—disputes, hierarchies, and competition for mates and influence. Communication evolved to meet the challenge. Voices sharpened, gestures grew more nuanced, and bonds deepened.

In this colder, more fragmented world, nature was selecting for social intelligence. The climate was no longer just shaping forests and rivers—it was sculpting minds.

Social Bonds Strengthen in a Harder World

As the great forests gave way to scattered trees and open plains, group life became indispensable. Mothers, once solitary, began raising their young together. The safety of the next generation—nursing, guarding, and grooming—became a shared responsibility.

Some individuals took on sentinel roles, climbing to high vantage points to watch for predators. One call, one cry, and the entire group could react. Vigilance became a collective defense. Groups didn't just protect; they taught. Elders remembered where water pooled after the rains or which trees bore fruit in the dry season, and the young learned through observation.

This was a new kind of inheritance—not written in DNA but passed through memory and teaching. Alliances formed, sometimes for protection, sometimes for power. Social navigation became a survival skill. Connection was no longer just comfort—it was currency.

With complexity came refined communication. Alarm calls began to take on specific meanings—one for eagle, another for leopard, another for snake. These calls represented ideas, not just sounds, a leap toward symbolic thought. Those who could learn, remember, comfort, persuade, warn, or soothe lived longer and passed on their genes.

Over time, this created a pre-linguistic toolkit—a set of sounds, gestures, and facial expressions bridging minds without words. It was not yet language, but it was the architecture of thought made visible—the foundation for stories yet to be told.

10 PARALLEL STRATEGIES— HERDS, PACKS, AND MINDS

"Man is by nature a social animal."
— **Aristotle**
(Philosopher)

Adapting to Each Other

Over the last few chapters, we've witnessed life's remarkable resilience through catastrophe—how mass extinctions became doorways to new possibilities, and how evolution repeatedly rewrote the story of biology. From the ashes of the Great Dying came dinosaurs. From the crater of cosmic impact came the age of mammals.

But among mammals, something novel was stirring. These warm-blooded survivors didn't just adapt to new environments—they began to adapt to each other. Social bonds became survival strategies. And cooperation became the equal of competition. From the complex dance of group living, consciousness itself began to emerge.

This is the story of how evolution discovered three parallel paths to survival—and how one of those paths led to minds capable of understanding their own existence.

The Mathematics of Survival

In the dangerous world that emerged after the dinosaurs' demise, individual mammals faced a stark calculation. A lone creature might find food more easily, avoid territorial disputes, and never have to share resources. But it also faced every predator alone, bore the full burden of finding food and shelter, and had no allies when crisis struck.

Evolution began experimenting with an alternative: the power of numbers. But not all social solutions are created equal. Across different environments and challenges, three distinct strategies emerged—each a different answer to the question of how individuals could benefit from proximity to others.

Strategy One: The Democracy of Motion

On the vast grasslands where trees were scarce and hiding places few, prey animals discovered the simplest and most effective social strategy ever devised: the herd.

Zebras, wildebeest, gazelles, and countless other grazers learned that survival was a numbers game. A predator can only catch one animal at a time. In a group of a hundred, your odds of being the unlucky one are already a low 1%. But in a group of 1000, your probability of becoming prey drops to 0.1%.

Herds required almost no social intelligence. There were no leaders, no followers, no complex relationships to navigate. The rules were elegantly simple: stay close, move together, and when danger comes, hope someone else is slower, younger, older, or weaker.

Communication in herds was minimal—a few alarm calls, some basic body language, the simple physics of group movement. When one animal bolted, others followed not out of understanding but out of instinct. Panic became a survival tool, spreading through the group faster than any predator could strike.

Yet within this apparent simplicity lay a profound innovation. For the first time in evolutionary history, individual survival became tied to group behavior. The herd was greater than the sum of its parts—not through intelligence or planning, but through the sheer mathematics of diluted risk.

This strategy worked so well that it spread across continents and persists today virtually unchanged. On African savannas, in American prairies, across Asian steppes, herds still move in ancient rhythms, individual minds submerged in collective motion.

Strategy Two: The Architecture of Power

In more complex environments—forests, mountains, river valleys— the simple mathematics of herding wasn't enough. Predators could approach unseen. Prey could hide or fight back. Food sources were scattered and defendable. In these landscapes, evolution discovered a more sophisticated social strategy: the pack.

Wolves, lions, hyenas, and wild dogs developed societies built around hierarchy, cooperation, and coordinated action. These weren't formless crowds but structured organizations with roles, ranks, and rules.

In a wolf pack, every individual knew their place in the social order. The alpha pair led hunts, settled disputes, and typically had first access to food. Subordinate wolves had their own ranks, their own relationships, and their own ways of competing for status without destroying group cohesion.

This hierarchy wasn't arbitrary cruelty—it was organizational efficiency. When a hunt began, every wolf knew their role without lengthy discussion. When danger threatened, leadership was clear. When conflicts arose, established dominance relationships prevented them from escalating into group-destroying violence.

Pack animals developed more sophisticated communication than those animals that relied on herds. Body posture, facial expressions, vocalizations, and scent markings created a rich vocabulary of intention and emotion. A subtle growl could prevent a fight. A submissive gesture could defuse tension. A coordinated howl could rally the group for action.

Most importantly, pack animals began to demonstrate something fresh: they could coordinate their behavior not just in response to immediate threats, but in pursuit of long-term goals. A pride of lions could plan an ambush, with different individuals taking specific positions to trap prey. A wolf pack could pursue injured prey for miles, taking turns leading the chase.

This was no longer simple instinct. It was strategy, requiring everyone to understand not just their own role, but how that role fit into a larger plan.

Strategy Three: The Revolution of Recognition

But in the most complex habitats—dense forests, cliff faces, river deltas—neither herding nor pack hierarchies were sufficient. These environments demanded something extraordinary: the ability to recognize other individuals as unique beings with their own thoughts, feelings, and intentions.

Among elephants wandering African woodlands, dolphins navigating coral reefs, and primates swinging through forest canopies, evolution discovered its most sophisticated social strategy: the alliance of minds.

These animals formed what scientists call fission-fusion societies—fluid groups where individuals came together and separated based on changing circumstances. Unlike the rigid hierarchies of wolf packs or the anonymous crowds of herds, these societies required each individual to maintain ongoing relationships with dozens of others.

An elephant matriarch must remember where distant water-holes endure during the dry season, which ancient paths lead safely to them, and when powerful musth bulls might pose danger to her calves. A chimpanzee had to track the shifting alliances within the troop, knowing who could be trusted or needed assistance, and who might betray or offered a threat.

This social complexity drove the evolution of larger brains, particularly the regions responsible for memory, emotional processing, and what scientists call "theory of mind"—the ability to understand that others have their own thoughts and feelings different from your own.

For the first time in Earth's history, animals began to engage in behaviors that required imagination—the ability to respond to others' emotional states rather than just their behaviors, the ability to teach by recognizing what others didn't know. The ability to deceive, which depended on understanding what others believed.

A chimpanzee mother wouldn't just protect her infant from immediate threats—she would anticipate dangers the infant couldn't perceive and guide it away from them. An elephant would not just trumpet in alarm when detecting danger—she would consider which family members were vulnerable and position herself to protect them.

This was the birth of empathy—not as a moral choice, but as a survival strategy. Animals that could understand and respond to others' mental states formed stronger alliances, provided better care for their young, and navigated social conflicts more successfully.

The Emergence of Mind-Reading

Within these complex social groups, evolution stumbled upon something that would change the course of life on Earth: minds that could think about other minds.

When a dolphin rescued an injured companion, it wasn't just responding to distress calls—it was imagining the other's experience of pain and acting to alleviate it. When a chimpanzee deceived a rival by pretending to find food elsewhere, it was modeling the rival's beliefs and deliberately manipulating them.

This cognitive revolution required new kinds of neural circuits. Mirror neurons allowed individuals to internally simulate others' actions and emotions. Enhanced memory systems tracked the history of relationships over time. Expanded prefrontal cortices in brains enabled the complex calculations required to predict others' behavior.

The social brain wasn't just bigger than solitary brains—it was qualitatively different. It could hold multiple perspectives simultaneously, imagine hypothetical scenarios, and plan behaviors based on predicted reactions of others.

This mental sophistication enabled new forms of cooperation. Animals could work together not just in immediate response to circumstances, but in pursuit of shared long-term goals. They could form coalitions, make deals, and even engage in what could only be called primitive politics.

The Costs of Consciousness

But this social complexity came with profound costs. Large brains required enormous amounts of energy—up to 20% of total caloric intake in some species. Extended childhoods were necessary to wire these complex neural circuits, leaving young vulnerable for years.

Most challenging of all, these animals had to navigate increasingly intricate social relationships while still meeting the basic demands of survival. A chimpanzee had to remember dozens of individual relationships, track shifting alliances, anticipate rivals' moves, and maintain friendships—all while finding food, avoiding predators, and raising offspring.

The cognitive load was immense. Some individuals thrived in this complex environment, becoming skilled social navigators who lived long lives and raised many offspring. Others struggled with the demands, becoming social outcasts or failing to reproduce successfully.

Evolution was conducting a vast experiment: Could the benefits of social intelligence outweigh its costs? Could brains complex enough to understand other minds also be efficient enough to ensure survival?

The answer would determine whether consciousness was an evolutionary dead end or the next great leap in the complexity of life.

The Feedback Loop of Intelligence

As social groups became more sophisticated, they created selection pressures for even greater intelligence. Individuals who could better understand others' motivations gained advantages in forming alliances, avoiding conflicts, and securing resources.

This created what scientists call a "cognitive arms race"—each generation needed to be slightly more socially intelligent than the last to compete successfully. Brain size increased. Neural connectivity became more complex. Childhood periods lengthened to allow more learning.

But this arms race also created something entirely new: cultures. Successful behavioral innovations could spread through groups not just through genetic inheritance, but through learning and imitation. A clever tool-use technique discovered by one individual could be adopted by others and passed down to offspring.

For the first time in Earth's history, evolution wasn't limited to the slow pace of genetic change. Behavioral evolution could happen within single lifetimes, allowing rapid adaptation to new challenges and opportunities.

The Birth of Teaching

Perhaps most remarkably, some of these socially intelligent animals began to teach—deliberately passing knowledge from one generation to the next through patient demonstration and correction.

A chimpanzee mother would slow her movements when showing her infant how to crack nuts, exaggerating each step and waiting for the young one to attempt imitation. A dolphin would bring her calf to safe shallow waters and demonstrate hunting techniques, repeating the lessons until the skills were mastered.

This wasn't just learning by observation—it was intentional education, requiring the teacher to understand what the student did and didn't know, and to modify their behavior accordingly.

Teaching represented a profound cognitive leap: the ability to hold two minds in awareness simultaneously—your own and another's—and to deliberately transfer information between them. This capacity would prove essential for the next phase of evolution: the development of language, culture, and ultimately human civilization.

Three Paths, One Future

Looking across the mammalian world, we see three successful strategies for social life, each adapted to different environments and challenges. Herds offer safety through numbers with minimal cognitive cost. Packs provide coordination through hierarchy and specialization. Alliance-based societies enable flexibility through individual recognition and mind-reading.

All three strategies persist today, each successful in its own ecological niche. But only one path led toward the kind of consciousness that could eventually contemplate its own existence: the path of minds learning to understand other minds.

This wasn't inevitable. Natural selection might never have favored the leap to symbolic thought. Life could have remained at the level of herds or packs—strategies already well suited to the demands of survival and group living. The neural investment required for complex social intelligence was immense, and its payoff far from certain.

Yet in a few lineages—dolphins, elephants, some birds, and most importantly for our story, primates—evolution continued pushing the boundaries of social cognition. These animals developed not just intelligence, but meta-intelligence: awareness of their own awareness, and awareness of others' awareness. Many even pass the classic "mirror test," touching or inspecting a hidden mark on their bodies after seeing it only in a reflection—evidence that they recognize themselves as individuals.

The Foundation of Everything Human

The three social strategies we've explored set the stage for understanding human nature. We carry within us echoes of all three approaches: the herd instinct that makes us follow crowds, the pack mentality that creates hierarchies and tribal loyalties, and the alliance-building intelligence that enables empathy, friendship, and love.

Our politics often resembles pack dynamics, with clear leaders and followers, competition for status, and loyalty to group identity. Our economics sometimes functions like herd behavior, with individuals following collective movements in markets and trends. Our personal relationships draw on alliance-building intelligence, requiring us to understand and respond to others' thoughts and feelings.

Understanding these evolutionary roots doesn't diminish human achievement—it illuminates the profound transformation that occurred when social intelligence became sophisticated enough to generate language, culture, and technology. We are pack animals who learned to think like alliance-builders, herd creatures who developed the capacity for genuine friendship.

Looking Ahead

The social intelligence that emerged from these parallel evolutionary experiments created the foundation for everything distinctly human: language that could share complex thoughts, cultures that could accumulate knowledge across generations, and technologies that could extend our capabilities beyond biological limits.

In our next chapter, we will explore how this social intelligence continued to evolve in our primate ancestors, creating the cognitive foundations for symbolic thought, shared meaning, and eventually the capacity to believe in things that exist only in our collective imagination.

The journey from herd to pack to alliance was just the beginning. From alliance-building minds would emerge something unique in the history of life: consciousness that could create its own realities through the power of shared belief.

A Reflection on Togetherness

Every human being alive today carries within their neural circuits the legacy of these ancient social experiments. When we feel the comfort of belonging to a group, we're experiencing the evolutionary wisdom of the herd. When we respond to leadership or compete for status, we're drawing on the pack brain that helped our ancestors coordinate action and survive challenges.

When we feel empathy for a stranger's suffering, offer help to someone in need, or form deep friendships based on mutual understanding, we're using the alliance-building intelligence that first emerged in complex social mammals millions of years ago.

But we've also transcended these ancient patterns. Through language and culture, we've created social bonds that span continents and generations. Through science and technology, we've built cooperation networks that no pack or alliance could achieve. Through art and story, we've shared mental worlds that exist nowhere else in nature.

We are social animals who became more than the sum of our social parts. We are the inheritors of billions of years of evolution's experiments in togetherness, and the creators of new forms of connection that the universe had never seen before.

Evolution discovered three parallel paths to survival through sociality: the anonymous safety of herds, the structured cooperation of packs, and the complex alliances built on individual recognition and empathy.

Each strategy succeeded in its own ecological niche, but only one path led toward the kind of consciousness that could eventually contemplate its own existence: the path of minds learning to understand other minds.

In our primate ancestors, this social intelligence was about to undergo a dramatic acceleration. Environmental pressures would force rapid changes in how groups lived, learned, and passed knowledge across generations. The individual conscious mind was just the beginning. The real adventure was in learning what minds could accomplish when they learned to think together.

11 THE EXPANDING MIND —FROM GROUP FEELING TO SHARED MEMORY

Consciousness bloomed not in isolation,
but in the space between minds
reaching toward each other.

The Space Between Minds

In our last chapter, we watched evolution experiment with three different approaches to social life: the anonymous safety of herds, the structured cooperation of packs, and the complex alliances built on individual recognition and empathy. We saw how the third path—minds learning to understand other minds—created the foundation for everything that would follow.

But recognition and empathy were just the beginning. As social intelligence deepened among our ancestors, something extraordinary began to emerge: the ability to share not just the present moment, but also memories, knowledge, and eventually dreams themselves.

This is the story of how consciousness learned to expand beyond the individual mind—how the capacity for shared experience became the foundation for culture, language, and the vast web of meaning that makes us human.

The Miocene Gardens

23 million years ago, during the warm Miocene epoch, much of Africa was a garden world that would have seemed almost magical to modern eyes. Vast forests stretched from coast to coast, broken by open woodlands and meandering rivers that reflected skies unmarked by contrails or the geometric boundaries of human civilization.

In this lush landscape, our distant ancestors—the early apes—lived lives of relative abundance. These primates were not yet humans, not even close. But they possessed something unprecedented in the long history of life: brains capable of genuine social intelligence that could recognize not just the presence of others, but the inner lives of others.

They lived in complex groups, where relationships mattered more than simple dominance. In these groups, individual recognition was the norm rather than the exception, and cooperation and conflict were both constant features of daily existence. An early ape mother didn't just protect her infant. She taught it which fruits were safe, which branches would hold its weight, and which calls meant danger or opportunity.

Perhaps most remarkably, these apes began to demonstrate something that had never existed before in the four-billion-year history of life: cultural transmission that could accumulate across generations. Useful behaviors could now spread through groups not by genetic inheritance—evolution's traditional method—but through observation, imitation, and practice. Knowledge became portable, moving from mind to mind like a beneficial contagion of wisdom.

This was consciousness preparing for its next great awakening—the realization that survival could be secured not just through sharper tools or stronger bodies, but through the invisible bridges of understanding and shared meaning.

The First Cultural Memory

What emerged from the social learning of observation, imitation, and practice was technology itself—the passing down of methods to shape the world. This was something revolutionary in the cosmos: group memory that transcended individual mortality. A technique discovered by a grandmother could still be used by her great-grandchildren, even though the original innovator had long since returned to the elements from which she came.

This wasn't just learning—it was cultural evolution, operating at speeds that genetic evolution could never match. While biological adaptations required thousands of generations to spread through populations, beneficial cultural behaviors could be adopted within years or even months.

The implications were staggering for the future of consciousness. For the first time in Earth's history, a species could adapt to new challenges not through the slow process of natural selection acting on random mutations, but through a rapid sharing of solutions between individuals and across generations.

A group that learned to use tools gained immediate advantages over those that didn't. A community that developed better communication could coordinate more effectively. Populations that accumulated useful knowledge became more successful than those that forgot their discoveries with each generation.

Evolution had stumbled upon something profound: minds connected by culture could evolve faster than minds isolated by individual experience. The universe was learning that consciousness worked best not in isolation, but in communion with other consciousness.

The Pressure of Change

This environmental challenge created intense pressure for innovation. Groups that could adapt quickly to changing conditions survived. Those that couldn't, faced extinction.

It was in this crucible of climate change that our lineage—the hominins—first appeared. The earliest human ancestors were still ape-like in many ways, but they possessed one crucial advantage over other primates: an enhanced capacity for cultural learning and memory.

Standing Tall, Thinking Deep

One of the first great adaptations our ancestors made was learning to walk upright. As forests thinned into open savannas nearly six million years ago, standing tall offered survival advantages: it freed the hands to carry food and infants, allowed the eyes to scan horizons for danger, and reduced the body's exposure to the harsh sun. Over millions of years, bones and muscles slowly reshaped—the pelvis broadened, the spine curved for balance, and the feet lost their grasping toes in favor of firm arches for forward motion. With hands no longer bound to the ground, our ancestors were free to shape tools, cradle fire, and eventually gesture meaning. Bipedalism was not an overnight change, but a transformation written into anatomy over vast stretches of time, the patient work of evolution shaping every step toward imagining what lay beyond the next hill.

But the real revolution was happening inside their skulls. As environmental pressures intensified, natural selection favored individuals with larger brains, better memories, and enhanced abilities to learn from others. These early hominins—species like *Ardipithecus* and later *Australopithecus*—not only moved more efficiently than their ape-like relatives but also began living in groups where cultural knowledge became essential for survival. They learned to remember the locations of water and food across a changing landscape, to shape stone into useful edges, and to navigate the fragile bonds of cooperation and conflict within their bands.

Evolutionary transition of early hominins: Ardipithecus (-4.4–5.8 million years ago), Australopithecus (-3–4 million years ago), Homo habilis (-2.1–2.4 million years ago), Homo ergaster (-1.9–1.4 million years ago), and Homo erectus (-1.9 million–110,000 years ago). Each stage shows key adaptations from ape-like ancestors to upright, tool-using humans with expanding brains and increasingly modern body proportions.

As these demands grew, so did the brain itself. The regions responsible for memory, planning, and social awareness began to expand. Neural connections multiplied, forming intricate networks of communication and recall. The cortex—the outer layer of the brain responsible for complex thought—grew thicker and more elaborate. With each generation, biology and culture reinforced one another, setting the stage for a species whose survival would depend as much on shared memory and imagination as on muscle and bone.

But perhaps most importantly, these early human ancestors developed something new: the ability to intentionally share complex information with others.

The Art of Tool Making

Around two and a half million years ago, our ancestors began making the first stone tools—carefully crafted implements that required skill, practice, and knowledge.

Making a proper stone tool isn't intuitive. It requires specific understanding, such as the properties of different rocks, knowing how to strike them at precise angles, and recognizing good rock edges from poor ones. Most crucially, it requires learning from someone who already knows how to do it.

Archaeological evidence suggests that our ancestors didn't just learn toolmaking through trial and error—they taught it deliberately. Adults showed children how to select proper stones, how to hold them correctly, and how to strike them with the right force and angle.

This was teaching in the true sense: intentional transfer of knowledge from one mind to another. It required the teacher to understand what the student didn't know and to modify their own behavior to fill that gap.

Teaching represented a cognitive revolution. It required sophisticated theory of mind—the ability to model another person's knowledge state and adjust your actions accordingly. It demanded patience, empathy, and the capacity to imagine how your own hard-won knowledge could benefit someone else.

The Expanding Circle

As cultural transmission became more sophisticated, the effective size of everyone's mind began to expand. A single person could now draw on the accumulated knowledge of their entire group and potentially groups they had never met.

A young hominin learning to make tools wasn't just accessing their own trial-and-error experience—they were tapping into generations of refinement and innovation. The techniques they learned had been tested by countless individuals, improved through countless iterations, and passed down through countless acts of teaching and learning.

This created what anthropologists call "the ratchet effect." Unlike other animals, who might discover useful behaviors but then lose them again when key individuals died, human cultural evolution could build progressively on previous innovations.

Each generation could start where the previous one left off, adding their own improvements to an ever-growing toolkit of knowledge and technique. Culture became a vast external memory system, like a hard drive, storing information that no individual brain could hold alone.

Homo Erectus: Minds on the Move

By two million years ago, this cultural evolution had produced a remarkable hominin: *Homo erectus*. With brains nearly twice the size of *Australopithecus* and their early Homo relatives such as *Homo habilis*, these early humans possessed unprecedented cognitive abilities. They crafted sophisticated tools that remained virtually unchanged for over a million years—a testament to their effectiveness. They controlled fire, built shelters, and organized cooperative hunts of large game. Most remarkably, they began to migrate out of Africa, spreading across Asia and establishing populations in habitats their ancestors had never seen.

This expansion was only possible because of their enhanced cultural learning abilities. When Homo erectus groups encountered new environments—different climates, unfamiliar plants and animals, novel challenges—they could adapt through cultural innovations that might spread within decades.

They learned to exploit new food sources, develop shelter appropriate to different climates, and modify their tool use for local conditions. Each successful adaptation could be shared within the group and passed to offspring, allowing rapid colonization of diverse environments.

The Social Brain Revolution

Yet the most important changes were happening not in their tools or their environments, but in their minds. Homo erectus lived in larger, more complex groups than their predecessors. They had to navigate relationships with dozens of individuals, remember complex social histories, and coordinate activities across extended networks.

This social complexity drove continued brain evolution. The prefrontal cortex—responsible for planning, decision-making, and social cognition—expanded dramatically. The temporal lobes, crucial for memory and language processing, grew larger and more interconnected.

Most importantly, the neural circuits responsible for empathy and theory of mind became more sophisticated. Homo erectus individuals could not only understand that others had different thoughts and feelings—they could predict what those thoughts and feelings might be in various situations.

This enhanced social intelligence enabled new forms of cooperation. Groups could plan complex activities requiring different individuals to adopt specific roles at specific times. They could negotiate conflicts through subtle social signals rather than violent confrontation. They could form alliances that transcended immediate kinship relationships.

The Dawn of Symbolic Thought

Scattered evidence suggests that Homo erectus may have taken the first tentative steps toward symbolic thought—the ability to use one thing to represent another. Some archaeological sites contain objects that seem to have been collected for their aesthetic, rather than pragmatic, value—objects like unusually shaped stones, and fossils or crystals.

While we can't know for certain what these objects meant to their collectors, they hint at a cognitive revolution: minds beginning to see significance beyond immediate function, to recognize patterns and meanings that transcended the purely practical.

This capacity for symbolic thought would prove crucial for everything that followed. It was the foundation for language, art, religion, and ultimately the vast web of shared meanings that characterizes human culture.

The Memory of Fire

Around this time, our ancestors made one of the most important discoveries in human history: the controlled use of fire. Fire created a focal point for group gatherings after dark. For the first time in our evolutionary history, our ancestors could extend their social interactions beyond daylight hours. Around the fire, they could share food, maintain tools, tend to the injured, and share knowledge. So, fire was more than just a tool—it was a social catalyst that transformed the very structure of human communities.

In the flickering light of those ancient fires, culture began to take on new dimensions. Without spoken language as we know it today, our ancestors nonetheless found ways to communicate complex information through gesture, demonstration, and perhaps simple vocalizations.

They shared memories of successful hunts, warnings about dangerous animals, knowledge of seasonal patterns and water sources. The fire became a library without books, a school without classrooms, a theater where the drama of daily life was reviewed, analyzed, and remembered.

The Network Effect

As groups became larger and more interconnected, the benefits of cultural learning multiplied. An innovation discovered by one group could spread to neighboring communities through trade, intermarriage, or simple observation.

This created network effects that accelerated cultural evolution. The more groups that shared knowledge, the faster beneficial innovations could spread. The larger the network of cultural transmission, the more diverse the pool of potential innovations.

Our ancestors were becoming part of something unmatched in the history of life: a species-wide learning network where discoveries made by individuals could benefit the entire population. They were laying the groundwork for the explosive cultural evolution that would eventually produce language, agriculture, cities, and science.

Looking Forward

The expansion of mind that began with our earliest ancestors was approaching a critical threshold. Soon, it would learn to create shared realities that existed nowhere but in the space between minds reaching toward each other in understanding. Enhanced social intelligence had created the capacity for cultural transmission. Cultural learning had accelerated adaptation to new environments and challenges. And symbolic thought was beginning to emerge.

All the pieces were in place for the next great leap: the development of true language and the explosion of human creativity that would follow. In our next chapter, we will witness this transformation, when our ancestors learned to do more than merely share knowledge, but instead share entire worlds of meaning through the power of spoken words.

A Reflection on Shared Knowing

When we teach a child to ride a bicycle, comfort a friend in grief, or collaborate on a complex project, we are continuing a tradition that began millions of years ago, with our earliest ancestors learning to share knowledge around flickering fires.

Every act of teaching, every moment of empathy, every successful collaboration draws on neural circuits that were shaped by countless generations of individuals who survived because they could understand and work with others.

We are not isolated minds trapped in separate skulls—we are participants in a vast, ongoing conversation that spans generations and continents. Our thoughts are shaped by ideas that originated in minds we will never meet. Our innovations build on discoveries made by people whose names are lost to history.

The expansion of mind that began in the Miocene continues today every time we connect with another person's thoughts, feelings, or experiences. We are living embodiments of consciousness that has learned to transcend its individual boundary, and of minds discovering that their greatest power lies not in isolation, but in the magical space where understanding passes from one awareness to another.

In this sense, consciousness was never really individual at all—it was always a collaborative creation, a shared project of minds working together to make sense of existence itself.

From shared memory to shared meaning, the next step is inevitable: the stories we tell begin to showcase how we shape identity, loyalty, and belief.

12 THE STORIES WE TELL OURSELVES—HOW BELIEF BECAME IDENTITY

*"It is far better to grasp the universe as it really is
than to persist in delusion,
however satisfying and reassuring."*
— **Carl Sagan**

The First Technology
of the Impossible

In our last chapter, we witnessed consciousness learning to expand beyond individual minds. We followed our ancestors as they developed the capacity for cultural transmission—knowledge that could be shared, accumulated, and passed down through generations. Around the controlled fires of Homo erectus, we saw the first glimmers of symbolic thought emerging, as communities gathered after dark to share not just food and warmth, but memories, warnings, and the accumulated wisdom of their groups.

These early humans had learned that minds connected by culture could evolve faster than minds isolated by individual experience. From tool-making techniques to seasonal knowledge, information became portable, moving from mind to mind like a beneficial contagion of wisdom.

But something even more profound was stirring around those ancient fires. Our ancestors weren't just sharing practical knowledge—they were beginning to create shared meanings that transcended immediate survival needs. They were learning to believe collectively in things that existed nowhere except in their imagination.

This was perhaps the most consequential invention in human history. Not fire, tools, or agriculture—but the capacity to believe collectively in stories that existed nowhere except in our shared imagination. Around those ancient fires, our ancestors discovered they could conjure entire worlds with nothing but words and faith. They could make the invisible feel real, the impossible seem inevitable, and the meaningless overflow with purpose.

It was humanity's first technology for conquering the impossible —challenges no individual could face alone, but that became achievable when hundreds of thousands acted together under a common belief. Over time, this shared imagination would evolve from a simple survival tool into a sacred identity—our greatest strength in uniting vast communities, and our most dangerous vulnerability when those beliefs collided.

The Architecture of Survival

To understand how we became a species of believers, we must first recognize what belief accomplished for our ancestors. It wasn't an accident or a mistake—belief systems emerged because they solved fundamental survival problems that pure logic couldn't touch.

Picture the world our ancestors inhabited: a realm of unpredictable dangers where droughts could last for years, where children died without warning, where hunters ventured out and never returned. The human mind, with its powerful hunger for patterns, desperately wanted to make sense of these chaotic forces. Why did the rains fail? Why did some tribes flourish while others vanished? Why did the same spirits that could bring life also bring death?

Belief systems provided answers where none existed. They transformed random suffering into meaningful trials, inexplicable events into the actions of gods and spirits. This didn't make the dangers less real, but it made them psychologically manageable. A tribe that could maintain hope and purpose in the face of catastrophe was more likely to survive than one that collapsed into despair when the universe seemed indifferent to their struggles.

But belief accomplished something even more remarkable. As human groups grew beyond the intimate bands where everyone knew everyone else, they faced an unprecedented challenge: how to maintain trust and cooperation among hundreds or thousands of individuals who were essentially strangers to each other.

Here, belief revealed its true genius. Shared stories created what we now call "imagined communities"—groups bonded not by blood or personal acquaintance, but by common faith in the same invisible realities. If everyone in the tribe believed in the same ancestors, feared the same spirits, and followed the same moral codes, they could cooperate as if they were family even when they had never met before.

A trader from across the valley could be trusted if he worshipped the same gods. A warrior could be counted on if he feared the same punishments in the afterlife. An entire civilization could coordinate its actions through shared belief in laws that existed only in their collective imagination.

The Natural Selection of Stories

But how did simple stories transform into sacred truths that people would die to defend? The answer lies hidden in the deep currents of time, where survival itself became the ultimate judge of which beliefs would endure, and which would fade into oblivion.

Over thousands of years, human groups unconsciously tested different belief systems against the harsh realities of existence. Like species competing for resources, stories competed for the most precious resource of all: human attention and devotion. Tribes with beliefs that encouraged cooperation, preserved crucial knowledge, and maintained group unity tended to flourish. Those with beliefs that promoted destructive behaviors or social chaos were more likely to fail and disappear from history.

This created a natural selection process among ideas—not survival of the truest beliefs, but persistence of the fittest, most adaptive ones. The stories that helped humans survive became the stories that humans believed.

When a tribe's rituals seemed to predict good hunting, when their ceremonies appeared to bring rain, when their moral codes prevented destructive conflicts, those beliefs gained credibility. Success reinforced faith, and faith motivated behavior that often led to more success. The fact that these connections might have been coincidental didn't matter—the correlation was enough to strengthen conviction.

Even when entire systems of belief collapsed, they did not vanish without leaving a trace. Fragments of myths, rituals, and symbols from unsuccessful traditions often found new life inside emerging ones.

A defeated tribe's gods might be reimagined as saints or spirits, their ceremonies woven into the festivals of the victors, their moral lessons retold in a different language or story. In this way, even failed belief systems contributed genetic-like material to the evolving lineage of human imagination. The most resonant parts—rituals that bound communities, symbols that inspired awe, codes that promoted survival—could be carried forward, embedded within a new framework of meaning. Thus, cultural evolution recycled useful ideas, ensuring that no story was ever lost entirely, only reshaped and repurposed to serve another generation's needs.

Gradually, the most successful beliefs became something more than useful stories. They became sacred and elevated beyond the realm of ordinary evidence and protected from criticism by powerful taboos. Once a story became sacred, it could motivate

extraordinary sacrifice and cooperation. People would die for sacred
beliefs in ways they never would for
mere practical arrangements.

The Chemistry of Faith

What we now understand, thanks to
modern explorations of the brain, is
that believing feels fundamentally
different from merely thinking.
When we hold strong beliefs—
whether religious, political, or
cultural—our neural networks
release the same chemicals associ-

An ancient gathering around
firelight — early humans share
stories, myths, and lessons that
bound their communities together.
These oral traditions became the
foundation of belief systems,
turning survival knowledge into
shared meaning.

ated with love, safety, and profound connection to something
greater than ourselves.

Shared religious practices trigger the release of oxytocin, the
same hormone that bonds parents to children and lovers to each
other. Believing together literally makes people feel connected at
the deepest biochemical level. When we encounter information
that confirms our existing beliefs, our brains release dopamine—
creating a warm flood of pleasure that makes confirming evidence
feel rewarding while contradictory information feels threatening
and unpleasant.

Repetitive religious activities—like chanting, singing, dancing,
or prayer—can trigger endorphins that create feelings of euphoria
and transcendence. Meanwhile, when core beliefs are challenged,
the brain's ancient alarm systems activate, releasing stress hormones
that make belief challenges feel like physical attacks on our very
survival.

This neurochemistry explains why belief feels so different from
opinion. Beliefs aren't just ideas we think are true—they're
emotional and social states that involve our entire being. Chal-

lenging someone's beliefs doesn't feel to them like correcting a factual error; it feels like threatening their identity, their community, and their deepest sense of safety in an uncertain world.

The First Sacred Stories

The earliest belief systems we can glimpse through the archaeological record shared remarkable similarities across vastly different cultures. Similarities that reveal their adaptive functions with startling clarity.

Nearly all early human societies developed beliefs about deceased ancestors continuing to exist after death, watching over the living while demanding respect and proper behavior. These ancestor beliefs served multiple survival functions simultaneously: they encouraged respect for elders' accumulated wisdom, promoted behaviors that would make ancestors "proud" (typically actions benefiting the group), and provided psychological comfort in the face of loss.

Early humans also populated their world with invisible beings associated with natural forces—river spirits whose moods encoded crucial information about seasonal flooding, forest entities whose demands reinforced sustainable hunting practices, weather gods who provided frameworks for understanding and responding to climate patterns. These beliefs didn't just create meaning; they preserved life-saving environmental knowledge within memorable, emotionally compelling stories.

Most cultures developed special individuals—shamans, medicine people, spiritual leaders—who could communicate with the invisible world. These figures served as living libraries of cultural knowledge, psychological healers, and social coordinators whose spiritual authority allowed them to make decisions and resolve disputes that might otherwise fragment communities.

Sacred places and objects provided tangible anchors for abstract beliefs. There were caves that marked important water sources, mountain peaks that signaled seasonal changes, and special tools and ornaments that preserved crucial cultural information through their symbolic associations. These physical touchstones gave invisible beliefs visible form, creating shared reference points that strengthened group identity across generations.

The genius of these early belief systems was how they encoded practical survival information within emotionally compelling narratives. The Golden Rule appears in virtually every human culture, not because ancient peoples discovered universal moral truth, but because groups that internalized reciprocal altruism as a sacred principle consistently out competed those that didn't. Dietary restrictions that seemed arbitrary often prevented disease. Sexual taboos maintained social stability. Ritual purification requirements limited the spread of infections.

When moral principles were commands from gods or ancestors, they carried an authority that no human leader could match or safely challenge.

When Stories Became Institutions

As human societies expanded beyond the scale of localized tribes, belief systems evolved from informal traditions into organized religions with formal structures, professional priesthoods, and systematic doctrines. This transformation amplified both the power and the dangers of collective belief.

The development of writing marked a cosmic moment in the evolution of belief. For the first time in the universe's history, stories could exist independent of the minds that created them. Ideas could outlive their authors. Beliefs could be preserved with almost perfect accuracy across generations.

Written scriptures enabled religious knowledge to spread farther and last longer than any oral tradition ever could, but they also introduced something that fluid oral traditions had largely avoided—the concept of orthodoxy, or "correct beliefs" that could be precisely defined and rigidly enforced.

When beliefs existed only in spoken form, they naturally

evolved with each telling, adapted to local circumstances and shaped by the wisdom of each storyteller. This flexibility had been one of belief's greatest survival advantages. Written scriptures changed this dynamic forever, creating official versions of truth preserved in unchanging form. Innovation became heresy. Deviation became blasphemy. The adaptive flexibility that had made early belief systems so powerful began to calcify into rigid doctrine.

As civilizations grew more complex, religious and political power began to merge like two streams flowing into a single river. Rulers discovered that claiming divine mandate gave their authority a sacred quality that mere human power could never achieve. Religious leaders found that political alliance provided protection, resources, and the ability to influence entire societies.

Sacred stories that had once served the spiritual needs of communities began to serve the political needs of empires. Doctrines were shaped to support existing power structures. Religious hierarchies mirrored social hierarchies. The gods themselves began to resemble earthly kings, ruling from celestial palaces with all the pomp and circumstance of human courts.

This wasn't corruption of pure religion by worldly politics—it was the inevitable evolution of belief systems operating at civilizational scales. When stories needed to organize millions of people across vast territories, they inevitably became entangled with political power. The result became belief systems that were more powerful than any that had come before. Yet they were also more dangerous. When religious conviction merged with political authority, the consequences of believing differently became a matter of life and death.

Belief as Tribal Identity

Perhaps most significantly, the institutionalization of belief changed humanity's relationship with the sacred itself. In early belief systems, the sacred had been immediate and personally experienced in dreams, visions, natural phenomena, and community rituals.

Organized religions created intermediaries between believers and the divine, requiring specialized knowledge and professional guidance.

More importantly, religious institutions began to function not just as spiritual systems but as tribal identities. To belong to a particular faith meant more than sharing certain beliefs—it meant belonging to a distinct community with its own customs, laws, and loyalties. Religious identity became a form of extended kinship, where fellow believers were treated as family members while those who believed differently were viewed with suspicion.

This tribal aspect served important functions, creating bonds of mutual obligation that could span vast distances, and providing social support systems for believers facing hardship. But religious tribalism also created new forms of conflict. Differences in belief became sources of tension between communities. Sacred differences could motivate violence that had little to do with actual theology and everything to do with group identity and competition.

As organized religions spread across the globe they inevitably came into conflict, as each system claimed exclusive access to ultimate truth. Religious wars became some of the most violent conflicts in human history, revealing a darker potential within belief systems, which had originally evolved to promote cooperation. When extended beyond their original communities, religious beliefs could become sources of division rather than unity.

The Secular Transformation

As the destructive potential of religious conflict became undeniable, some human societies began experimenting with a radical idea. What if communities were organized around shared principles rather than shared gods? Beginning in the late 17th century, and flourishing throughout the 18th century, a revolutionary intellectual movement emerged that would forever change how humans thought about authority, knowledge, and governance.

The Enlightenment—spanning from roughly 1685 to 1800—represented humanity's first systematic attempt to build civilization on reason rather than revelation. Thinkers like John Locke, Voltaire, and Rousseau dared to ask questions that had been forbidden for centuries: What if governmental authority came not from divine mandate but from the consent of the governed? What if human beings possessed inherent rights that no king or pope could legitimately violate? What if truth could be discovered through observation and logic rather than ancient texts and religious doctrine?

These ideas emerged from the ashes of Europe's devastating religious wars, which had torn the continent apart for over a century. The Thirty Years' War alone had killed roughly one-third of Central Europe's population, all because of competing versions of Christian truth. Faced with such carnage, a new generation of philosophers concluded that organizing society around competing religious certainties was a recipe for endless conflict.

The Enlightenment introduced revolutionary concepts about natural rights, such as the radical notion that humans possessed inherent dignity and freedom simply by virtue of being human, not because any authority granted these privileges. The movement promoted rational governance based on evidence and debate rather than tradition and blind obedience. Most audaciously, it championed religious tolerance—the idea that different faiths could coexist peacefully within the same society, each free to worship according to conscience.

These principles challenged the traditional authority of religious institutions in ways that seemed almost inconceivable to previous generations. For the first time in human history, large groups began to imagine societies bound together by evidence rather than faith, by reason rather than revelation.

But this secular awakening revealed something profound about human psychology: eliminating religious authority didn't eliminate the human need for shared beliefs and collective identities. Instead, it redirected those needs toward new objects, creating secular equivalents of the psychological functions that religion had always served.

The French revolutionaries of 1789-1799 didn't just overthrow the monarchy and aristocracy in pursuit of liberty, equality, and fraternity—they systematically created new secular belief systems to replace the religious foundations of the old order. They established festivals celebrating Reason and the Supreme Being to replace Catholic holidays, designed new symbols like the tricolor flag and Marianne to replace crosses and royal emblems, and invented civic rituals and ceremonies to replace masses and religious observances. Even as they executed King Louis XVI and dismantled centuries of royal tradition, they understood intuitively that humans needed shared sacred stories and collective rituals to maintain social cohesion. The communist movements of the twentieth century functioned similarly, demanding total commitment and promising earthly salvation through historical inevitability rather than divine intervention.

Nationalism emerged as perhaps the most powerful secular religion, complete with sacred symbols (flags), holy texts (constitutions), martyrs (war heroes), and elaborate rituals (patriotic ceremonies). Political ideologies like socialism, fascism, and capitalism developed their own forms of orthodoxy and heresy, inspiring zealous devotion that rivaled any traditional faith. Economic systems became objects of quasi-religious belief, with their own prophets, sacred principles, and promised redemptions.

Even in the most secular societies, the psychological functions once served by religious belief found new expressions. People still needed to belong to something larger than themselves, still needed shared stories to make sense of existence, still needed group identities to navigate an uncertain world. The content had changed—reason replaced revelation, nation replaced church, ideology replaced theology—but the underlying human needs remained constant.

The secular age had not transcended belief—it had simply provided other things to believe in, often with the same intensity and exclusivity that had once characterized religious faith. The same psychological mechanisms that had made religious belief both powerful and dangerous now played out in political movements, national identities, and ideological commitments.

This transformation revealed a crucial insight: the capacity for belief isn't a primitive remnant of humanity's religious past, but a fundamental feature of human psychology that adapts to whatever content a culture provides. Secularization didn't eliminate the tribal, emotional, and meaning-making aspects of human nature—it channeled them through new institutions and narratives that served the same psychological functions as their religious predecessors.

The Digital Age of Faith

In our contemporary world, the evolution of belief has entered an entirely new phase. Digital networks have created unrivaled possibilities for belief formation and transmission, while also exposing the ancient psychological mechanisms that make humans vulnerable to manipulation through shared stories.

Social media algorithms have become the new priests of belief, determining which ideas reach which minds based on engagement rather than truth. Echo chambers and filter bubbles recreate the tribal boundaries that once protected small groups but now fragment global society into incompatible realities.

False beliefs can spread across the planet in hours, carried by the same technological networks that spread genuine knowledge. Conspiracy theories bind groups together around shared narratives that serve the same psychological functions as ancient myths, providing meaning, identity, and explanation for suffering—even when those narratives contradict all available evidence.

Yet the same digital tools that enable the spread of misinformation also create immense opportunities for human connection across traditional boundaries. Virtual communities form around shared interests rather than shared geography. Global movements coordinate action across continents in real time. We are witnessing belief evolution in real time, accelerated by technology but driven by the same fundamental human needs that first sparked religious consciousness around ancient fires.

The Challenge of Conscious Belief

Despite centuries of scientific advancement and secular education, belief systems continue to evolve and thrive. Humans seem to need sacred stories, transcendent purposes, and group identities that go beyond immediate material concerns. Even the most scientifically literate individuals often hold beliefs that serve psychological rather than evidential functions.

This suggests that the capacity for belief isn't a primitive remnant that education can eliminate, but a fundamental feature of human psychology that serves important functions in navigating complex social and existential challenges. The question isn't whether humans will continue to believe—we almost certainly will. The question is whether we can develop forms of belief that promote cooperation rather than conflict, wisdom rather than dogma, and openness rather than fanaticism.

Understanding the evolutionary origins and psychological functions of belief systems creates new possibilities for conscious choice about what we believe and why. We can recognize that our desire for belonging, meaning, and shared identity is a legitimate human need, while being more thoughtful about how we fulfill those needs.

We can appreciate the social functions of belief systems while being more critical about their claims to absolute truth. We can participate in communities of shared values and purpose while remaining open to evidence and perspectives that challenge our assumptions. And we can hold beliefs firmly enough to guide our actions and commitments, while also holding them lightly enough to account for change when circumstances require.

The Stories We Need Now

As we face global challenges requiring unparalleled cooperation—like climate change, technological risks, and growing inequality—we may need to consciously evolve new belief systems adapted to our current circumstances.

We need stories that help us feel connected to all humans rather than just our immediate tribes. We need narratives about our shared cosmic origins, our common evolutionary heritage, and our mutual dependence on planetary systems. We need beliefs that motivate care for future generations, and expand our sense of family to include descendants not yet born.

We need methods to honor scientific knowledge that acknowledges both its power and its limitations, so we can develop the intellectual humility necessary to distinguish between what we know and what we assume. And finally, we need belief systems that can evolve as quickly as our circumstances while providing enough stability to offer meaning and guidance through turbulent times.

The Eternal Return

Throughout this exploration, we return to a fundamental truth: humans are meaning-makers in a cosmos that operates according to impersonal laws.

This creates both our greatest challenge and our greatest opportunity. The challenge is learning to distinguish between stories that expand our possibilities and those that constrain them, between beliefs that serve life and those that serve only themselves.

The opportunity is recognizing that we are not passive recipients of inherited beliefs, but rather active creators of the narratives that will shape humanity's future. We can choose stories of abundance over scarcity, cooperation over competition, and hope over fear. The universe may be indifferent to our narratives, but it is not indifferent to their consequences.

The stories we tell ourselves become the realities we create. In a world where human actions affect the entire planet, our capacity for conscious storytelling has never mattered more. Every human being alive today carries within their consciousness some version of the sacred stories that have sustained our species for millennia. Whether expressed through traditional religions, political ideologies, scientific worldviews, or personal philosophies, we all need narratives that connect our individual existence to something larger and more enduring than ourselves.

The task of our time is learning to hold these stories consciously rather than unconsciously—to recognize their psychological and social functions while being honest about their relationship to empirical truth. We can appreciate the wisdom embedded in ancient traditions while adapting that wisdom to contemporary challenges. We can celebrate the human capacity for meaning making while remaining humble about the limits of any particular meaning system.

Perhaps most importantly, we can recognize that the sacred stories of other people—even when they seem strange or wrong to us—serve the same fundamental human needs that our own stories serve. This recognition doesn't require us to accept all beliefs as equally valid, but it can help us approach differences with greater empathy and less hostility.

The age of unconscious belief is ending. The age of conscious myth-making has begun. And in the space between what is and what could be, we continue the ancient human tradition of conjuring meaning from the darkness, one shared story at a time.

13 ANCIENT MINDS IN A DIGITAL WORLD

"The real problem of humanity is the following: we have Paleolithic emotions, medieval institutions, and godlike technology."
— **E.O. Wilson**
(Biologist)

Ancient Software, Modern World

The previous chapter left us standing at the crossroads of human belief, where survival stories whispered around ancient fires grew into organized religions capable of uniting millions, and eventually into secular ideologies promising salvation on Earth. We saw how the same psychological mechanisms that once helped small tribes endure have been channeled through increasingly powerful institutions and technologies.

But that story didn't end with the rise of secular thinking or even the digital revolution. It continues today, playing out in real time across social media networks, political movements, and global crises. The same ancient brain that once helped our ancestors distinguish friend from foe now scrolls through Facebook, votes in elections, and decides whether to trust climate scientists.

With 200,000-year-old neural software, we try to navigate a world that changes faster than our minds can adapt. And nowhere is this mismatch more visible—or more dangerous—than in how we form beliefs, choose leaders, and respond to information in our hyper-connected age.

This is the story of what happens when Stone Age psychology collides with Space Age problems.

The Tribal Pulse in the Digital Bloodstream

Every morning, millions of people wake up and immediately check their phones, scrolling through carefully curated feeds of information designed to capture and hold their attention. What they don't realize is that they're engaging in the most sophisticated tribal sorting system ever created—one that would be instantly recognizable to our cave-dwelling ancestors.

Social media algorithms don't just show us information; they show us information that confirms what we already believe. This technology connects us with people who think like us and gradually insulates us from perspectives that might challenge our worldview. This isn't accidental. It's how these systems are designed to work. Engagement drives profit, and nothing engages us quite like having our existing beliefs validated and our tribal loyalties reinforced.

Consider how political beliefs spread through social networks today. Information doesn't flow randomly—it moves along tribal lines with the same patterns that once carried gossip, warnings, and shared stories through ancient communities. Democrats share articles that confirm their suspicions about Republicans. Republicans circulate content that validates their concerns about Democrats. Each group develops its own version of reality, its own trusted sources, its own explanation for why the other side seems so wrong.

This creates what researchers call "epistemic tribes"—groups that don't just share political preferences but entire ways of knowing what's true. Members of different epistemic tribes can look at identical evidence and reach completely opposite conclusions, not because they're stupid or evil, but because they're filtering information through different tribal lenses that evolved to maintain group cohesion rather than seek objective truth.

The psychological rewards are the same as those that once bound hunting parties and war bands together. When someone in your political tribe shares information that confirms your beliefs, your brain releases dopamine—the same chemical reward that once reinforced successful cooperation in small groups. When you encounter information that contradicts your tribe's beliefs, your brain activates threat-detection systems—the same neural circuits that once identified dangerous outsiders.

We experience political disagreement not as a difference of opinion, but as an attack on our identity, our community, our very survival. And we respond accordingly—not with curiosity or nuance, but with the fierce loyalty and defensive aggression that once protected our ancestors from real physical threats.

The New Mythology Machine

But digital tribalism goes beyond just political polarization. It creates entirely new forms of mythology—modern belief systems that serve the same psychological functions as ancient stories but spread at the speed of light through global networks.

Conspiracy theories are perhaps the purest example of ancient storytelling adapted to modern media. They provide the same benefits that sacred stories once offered: they explain confusing events, identify enemies and allies, create meaning from chaos, and bind believers together in communities of shared understanding.

Consider how QAnon emerged and spread. It began with cryptic messages posted anonymously on internet forums, claiming to reveal hidden truths about powerful elites. These messages were deliberately vague and mysterious, requiring believers to interpret and decode their meaning—a process that made them feel like special initiates with access to sacred knowledge.

The stories that emerged had all the elements of classical mythology: a cosmic battle between good and evil, hidden knowledge revealed only to the faithful, a promised salvation when the truth would finally be revealed. Believers formed communities around these stories, developing their own language, rituals, and social hierarchies. They experienced the same sense of purpose, belonging, and meaning that humans have always derived from shared myths.

The fact that these stories contradicted massive amounts of evidence didn't diminish their power—it enhanced it. Just as ancient myths weren't meant to be scientific descriptions of reality, modern conspiracy theories aren't really about facts. They're about identity, community, and psychological needs that no amount of fact-checking can address.

Digital networks accelerate this process exponentially. A conspiracy theory that might once have taken years to spread through a community can now reach millions of people within hours. Social media algorithms, optimized for engagement rather than truth, actively promote content that generates strong emotional responses—exactly the kind of content that conspiracy theories provide.

Meanwhile, the same technologies that spread misinformation also create the illusion of research and verification. People can find thousands of websites, videos, and social media posts that support almost any belief, no matter how disconnected from reality. The abundance of information available online makes it easier, not harder, to maintain false beliefs.

The Algorithm as High Priest

In traditional societies, religious and political leaders served as gate-keepers of information, determining what stories their communities would hear and believe. In our digital age, that function has been largely taken over by algorithms—mathematical formulas that decide what information reaches which minds.

But unlike human leaders, algorithms don't make conscious decisions about truth or falsehood, benefit or harm. They optimize for simpler metrics: engagement, clicks, time spent on platform, shares, and comments. This creates a fundamental mismatch between what algorithms promote and what humans actually need.

Information that makes us angry, afraid, or outraged generates more engagement than information that makes us thoughtful, nuanced, or uncertain. Stories that confirm our existing beliefs feel more satisfying than facts that challenge our assumptions. Content that divides people into clear categories of good and evil spreads faster than content that acknowledges complexity and ambiguity.

The result is that our information environment becomes systematically biased toward content that triggers our most ancient psychological reactions—fear, anger, tribal loyalty, and the need for simple explanations in a complex world. We're not receiving a representative sample of available information; we're receiving information selected for its ability to activate our most primitive emotional responses.

This creates what researchers call "artificial scarcity" of nuanced, thoughtful information. There's no shortage of careful analysis, scientific research, and balanced reporting in the world—but algorithm-driven media systems don't prioritize content that requires reflection over content that generates immediate reaction.

The most engaged users—those who click, share, and comment most frequently—tend to have the strongest and most extreme views. Their behavior signals to algorithms that polarizing content is what audiences want, creating feedback loops that push public discourse toward ever-greater extremes.

Meanwhile, people with moderate views and nuanced positions become less visible in digital spaces, not because they don't exist, but because their behavior doesn't generate the signals that algorithms are designed to amplify. The result is a digital public square that makes society appear more divided and extreme than it is.

When Science Meets Tribal Loyalty

Perhaps nowhere is the collision between ancient psychology and modern challenges more visible than in how we respond to scientific information that contradicts our tribal loyalties.

Climate change presents a perfect case study. The scientific evidence for human-caused global warming is overwhelming—supported by virtually every climate scientist, confirmed by multiple independent lines of research, and backed by decades of increasingly precise data. Yet acceptance of climate science correlates strongly with political identity, especially in the United States.

This isn't because climate science is particularly complicated—most people can understand the basic greenhouse effect. It's because climate science has become associated with political tribes that have different values, different preferred solutions, and different views about the role of government in addressing collective problems.

For many conservatives, accepting climate science feels like accepting that government regulation is necessary, that individual freedom must be constrained for collective benefit, and that their preferred political leaders have been wrong about an important issue. The psychological cost of updating their beliefs is enormous—it requires not just changing their mind about scientific facts, but potentially changing their political identity, their social relationships, and their fundamental worldview.

The same dynamic plays out across numerous scientific issues. Vaccines become political when they're associated with government mandates. Evolution becomes controversial when it conflicts with religious identity. Even basic health measures during a pandemic become tribal markers when they're promoted by leaders of the opposing political party.

Our brains evolved to trust information from our own tribe and distrust information from outsiders, especially when that information threatens our group's status or requires costly behavioral changes. This made perfect sense when humans lived in small groups where tribal loyalty was essential for survival. It becomes actively dangerous when we need to coordinate responses to global challenges that require accurate information and collective action regardless of tribal boundaries.

The Challenge of Planetary Thinking

The ultimate test of our ability to transcend tribal psychology is climate change—a global problem that requires coordinated cooperation across all traditional boundaries of nation, culture, and ideology.

Climate change violates almost every assumption built into our evolved psychology. It's gradual rather than immediate, statistical rather than dramatic, global rather than local, and requires sacrificing short-term benefits for long-term survival. Worse, it requires trusting information from scientists we've never met and cooperating with people in distant countries who may seem like competitors rather than allies.

Our tribal instincts make us want to identify enemies we can fight and defeat. But climate change has no army to battle, no territory to defend, no clear victory conditions. It's a systems problem that requires systems thinking—the ability to understand complex interactions across multiple scales of space and time.

Consider how climate denial spreads and persists despite overwhelming scientific evidence. It's not primarily about the science—it's about identity, community, and psychological needs. Climate denial provides what all successful belief systems provide: a sense of group belonging, a way to resist outside authority, and stories that make a complex world feel simpler and more controllable.

This creates a tragic irony—the species that evolved the intelligence to understand global climate systems also evolved the psychological limitations that prevent us from responding appropriately to that understanding.

The Nationalism Trap

Nowhere is the conflict between tribal psychology and global needs more apparent than in the rise of nationalist movements around the world. As globalization creates economic uncertainty and cultural change, many people retreat to the safety of national identity—the modern equivalent of tribal belonging.

Nationalism appeals to deep psychological needs that evolution hardwired into our brains. It provides clear group boundaries (us versus them), shared stories about the past (national myths), common enemies (other nations or global elites), and simple solutions to complex problems (put our nation first).

The emotional rewards of nationalism are immediate and powerful. National symbols trigger the same pride and belonging that once came from tribal identity. National leaders become father or mother figures who promise to protect the group from outside threats. National myths provide meaning and purpose that transcend individual concerns.

But nationalism creates systematic obstacles to addressing global challenges. Climate change, pandemic response, economic stability, and technological governance all require international cooperation on scales that nationalism actively discourages. When nations prioritize their own short-term interests over collective long-term survival, everybody loses.

The most successful nationalist movements understand this psychological dynamic and exploit it deliberately. They identify outside groups to blame for internal problems, promise simple solutions to complex challenges, and create emotional narratives that make supporters feel special, threatened, and morally superior simultaneously.

The irony is that many of the problems that fuel nationalist resentment—economic inequality, cultural disruption, political powerlessness—are themselves caused by global forces that can only be addressed through international cooperation. Nationalism offers psychological comfort but prevents the collective action that could actually solve underlying problems.

The Illusion of Knowledge

The internet promised to democratize access to information and break down barriers between different communities and perspectives. Instead, it often creates the illusion of research while reinforcing existing beliefs.

Consider how people "research" controversial topics online today. They typically start with search terms that reflect their existing assumptions, click on results that confirm what they already think, and share findings with online communities that think similarly. This process feels like objective investigation but in reality, functions as confirmation bias on steroids.

The abundance of information online makes this worse, not better. For almost any belief—no matter how incorrect—it's possible to find websites, videos, studies, and expert opinions that support it. People can spend hours consuming content that makes them feel informed while becoming less informed about actual reality.

Social media compounds this problem by creating personalized information environments that gradually become more extreme over time. Each click, share, and like provides data that algorithms use to predict what content will generate more engagement. This creates feedback loops that push users toward increasingly radical versions of their existing beliefs.

The result is communities of people who share the same misinformation, reinforce each other's false beliefs, and develop increasing confidence in positions that are disconnected from evidence. They experience the social validation that comes from group agreement without the reality testing that comes from genuine debate with knowledgeable opponents.

This creates what researchers call "confident ignorance"—the state of being highly certain about beliefs that are factually incorrect. People in this state don't feel like they're lacking information; they feel like they have special access to truth that others are too ignorant or biased to recognize.

The Weaponization of Ancient Instincts

Perhaps most concerning is how bad actors have learned to exploit our evolutionary psychology deliberately, using our ancient instincts against us for political, economic, or ideological gain.

Foreign governments run sophisticated disinformation campaigns designed to amplify existing social divisions in their target countries. They don't need to create conflict from scratch—they just need to identify existing tribal tensions and amplify them through carefully crafted content that triggers strong emotional responses.

These operations succeed because they work with, rather than against, human psychology. They create fake social media accounts that purport to be members of different political tribes, sharing content designed to make each group more suspicious and angry toward each other. They promote conspiracy theories that make complex problems seem to have simple explanations and clear villains.

Domestic political actors use similar tactics, using focus groups and data analysis to identify the messages most likely to trigger fear, anger, and tribal loyalty among their target audiences. They craft narratives that make supporters feel threatened by outsiders and dependent on strong leaders for protection.

Commercial actors exploit these same psychological vulnerabilities for profit, creating business models based on capturing and holding human attention through emotional manipulation. The more time people spend engaged with their platforms, the more advertising revenue they generate—regardless of whether that engagement is making users happier, better informed, or more capable of functioning in society.

The result is an information environment where truth competes with lies, facts compete with feelings, and careful analysis competes with emotional manipulation—often on an unequal playing field where manipulation has systematic advantages because it aligns with rather than challenges our evolved psychological tendencies.

The Path Through Ancient Shadows

Yet understanding these challenges also points toward potential solutions. If we recognize that our problems stem from mismatches between ancient psychology and modern circumstances, we can design systems that work with rather than against human nature.

This might mean redesigning social media algorithms to prioritize accuracy over engagement, teaching media literacy in schools, creating institutions that facilitate constructive dialogue across tribal boundaries, and developing new forms of governance that channel competitive instincts toward collective benefit rather than zero-sum conflict.

Some promising developments are already emerging. Fact-checking organizations work to counter misinformation. Bridge-building groups bring together people from different political tribes to find common ground. Scientists and educators develop ways to communicate complex information that is less likely to trigger defensive reactions.

Most importantly, growing numbers of people are becoming aware of their own psychological limitations and actively working to overcome them. They seek out diverse perspectives, practice intellectual humility, and make decisions based on evidence rather than tribal loyalty.

This awareness itself may be the most important development. For the first time in human history, we have the scientific knowledge to understand our own cognitive biases and emotional responses. We can recognize when our ancient instincts are leading us astray and consciously choose different responses.

The Future of Human Judgment

As we face global challenges that require enormous cooperation and accurate information processing, the question isn't whether we can eliminate our evolutionary psychology. We can't, and we shouldn't try. These ancient instincts also bequeath us creativity, empathy, moral intuition, and the emotional bonds that make life meaningful.

The question is whether we can learn to recognize when our instincts serve us well and when they lead us astray. Can we develop the wisdom to trust our emotions in personal relationships while relying on evidence for factual questions? Can we enjoy the comfort of tribal belonging while recognizing our common humanity? Can we honor our cultural traditions while remaining open to new information?

The stakes could not be higher. The same technologies that amplify our ancient biases also provide us colossal power to shape the future of our species and our planet. Will artificial intelligence enhance human flourishing or concentrate power among elites? Will climate change lead to cooperation or conflict? Will global challenges bring out humanity's best or worst impulses? The outcomes of all these choices will be determined by how well we manage the interaction between ancient psychology and modern technology.

We are the first generation in human history to understand both our cosmic origins and our psychological limitations. We are also the first generation with the power to consciously direct our own evolution—not only biological evolution, but also the cultural and technological evolution that now shapes our future faster than genetic change.

In this chapter, we learned that belief gave us identity, but imagination alone could not cut wood, pierce hide or hold back the cold. Survival demanded something tangible—edges struck from stone, sparks coaxed into flame, seeds pressed into soil. These inventions were more than objects; they were extensions of the body and mind, shaping both our future and us. In the next chapter, we follow this other great current of our story—the rise of the maker —tracing how tools carried us from stone to steam, multiplying our power with every step.

14 MAN THE MAKER— FROM STONE TO STEAM

*"The stone age did not end because
we ran out of stones, but because
we invented something better."*
— **Ahmed Zaki Yamani**
(Former Saudi Oil Minister)

The Thread of Making

Before long, belief had bound scattered tribes into something greater than themselves. Myths and shared meanings allowed hundreds, even thousands, of strangers to act as one. Yet imagination alone was never enough. Stories could inspire, but survival required something solid in the hand—something to cut, scrape, hammer, or burn.

So, alongside the myths, another thread began weaving itself through human history: the thread of making. From the beginning, we were not just storytellers. We were *tool builders*.

A sharp stone, chipped just right, was the first external claw. A pointed stick was the first tooth that could pierce the hide of a beast larger than us. Each invention was not just an object but an extension of our own bodies. When our ancestors first struck stone against stone and held the edge that broke away, they were shaping more than rock—they were shaping the future.

Stone and Bone—The First Tools

Nearly two and a half million years ago, long before Homo sapiens, our ancestors struck flakes from riverbed stones and discovered the cutting edge. Oldowan tools, the earliest known stone toolmaking tradition used by early hominins, were simple, but they multiplied the power of bare hands. With them came marrow cracked from bone, roots split open, hides scraped clean.

Soon came hand axes—teardrop-shaped stones with sharp edges on both sides. They were more than tools; they were *teaching objects*. To make one required memory, practice, and imitation. Tools became part of our culture; knowledge carried not in DNA but in demonstration.

Now, each generation inherited techniques as well as genes. A child watched, copied, and perhaps surpassed. In this way, technology became its own form of inheritance—a second stream of evolution running alongside biology.

With tools, our reach extended. The clawless primate gained talons of flint. The frail hand grew strong enough to butcher an elephant.

Fire—The Captured Star

If stone was the first claw, fire was the first engine[PM1] of change. Somewhere in Africa, perhaps a million years ago, our ancestors dared to touch flame. At first it must have been taken, from lightning strikes, or from wildfires sweeping the savanna. But once tamed, fire became a revolution.

Around its light, the night retreated. Predators were pushed away from the edges, afraid. Meat, once tough and dangerous to digest, softened and gave up more calories. The very shape of our bodies changed: smaller jaws, smaller guts, larger brains.

Cooking was chemistry turned inward, unlocking the stored sun of plants and prey. With fire, winter lost some of its bite. With fire, dark caves became habitable homes.

But fire was more than a survival trick. It was a symbol, the first hint that humans might one day wield powers that belonged to the cosmos itself. We had captured a fragment of the Sun and made it ours.

And yet, in those first flames, one can already glimpse the shadow of today's crisis. For when we light fossil fuels, we are still burning the stored energy of ancient life. Fire gave us a future, but it also set us on a path that could burn it away.

Each invention carried the blueprint of the next. A chipped stone edge made it easier to carve wood into spears and handles. Fire hardened those spears, then softened metals that would become plows and blades. Farming tools multiplied the harvest, but surplus grain created a new problem: keeping track. Clay tokens became symbols, and symbols became writing—the first marks of history. With every step, the tool itself contained instructions for a more powerful one to follow. In this way, invention learned from itself, accelerating with each generation.

When Seeds Became Cities

Ten thousand years ago, humanity performed its boldest experiment: we planted seeds and waited.

Agriculture was a leap of faith. It tied us to the soil, to rivers and rains, to the cycles of sun and season. But in return, it yielded abundance. With fields came surplus. With surplus came villages. With villages, in time, the first cities.

Farming created stability, but also dependence. Drought meant famine. Pests meant ruin. Where a hunter could move on, a farmer was bound to the land. To manage risk, people began to store food, trade grain, invent laws, and record harvests in symbols that became writing.

Agriculture also transformed landscapes. Forests were felled, rivers redirected, wild grasses bent into wheat and rice. Humans were no longer just shaped by ecosystems—we had become shapers of them.

This tether to the land birthed civilization itself. Religion and politics grew alongside irrigation canals. Kings, priests, and armies arose to protect and control the surplus. Survival was no longer just biological. It was cultural, political, and technological.

Metal and Machines —The Age of Empires

If farming tied us to the soil, metal set us loose upon each other.

Bronze tools cut deeper. Iron swords conquered wider. With smelting and forging, empires could till vast fields, arm thousands of soldiers, or build ships to span seas. Civilization's reach widened, and with it, the human footprint.

Every breakthrough in metallurgy was also a breakthrough in power. The plow turned soil more efficiently, feeding armies. The sword and spear defended grain stores or plundered them from others. Where belief united, technology divided, setting civilizations on collision courses.

And yet, even here, invention followed the same thread: survival. Groups with stronger tools endured. Groups without them vanished or were absorbed. The race of invention became a race of survival itself; a contest of civilizations sharpened on anvils and whetstones.

Steam and Smoke—The Industrial Revolution

Then, in the 18th century, came the most radical shift since fire.

Coal from the Carboniferous age—basically buried sunlight—

was fed into furnaces. Out of these furnaces came steam, an invisible, unstoppable pressure that drove pistons and wheels.

The steam engine multiplied labor beyond anything muscle had ever achieved. Mills turned night into day. Railroads stitched continents together. Ships crossed oceans with cargoes unimagined.

For the first time, survival no longer depended on human strength or animal power. Machines, fueled by carbon, did the work of millions.

The result was an explosion of cities, factories, and populations. Medicine and sanitation extended life. Mechanization expanded production, and the world's human population soared. In just two centuries, the population grew from less than 900 million people to more than seven billion people.

But this miracle carried two hidden costs. Burning coal, oil, and gas released vast amounts of carbon dioxide, a greenhouse gas that traps heat and disrupts Earth's climate system. At the same time, those fuels were finite stores of energy—carbon locked away in ancient plants and seas for hundreds of millions of years. Each factory and train gained power in the present, but at the expense of a limited inheritance that cannot be replaced.

We had cracked open the vault of Earth's long history and spent its treasure in a heartbeat.

The Arc of Creation

Looking back, the pattern is unmistakable. Each link in the chain of invention followed from the last, each forged from the same instinct: to endure.

- A stone edge to cut meat.
- A flame to warm the night.
- A seed to feed a village.
- A plow, forge, and wheel to ramp up production.
- A furnace to burn the past itself.

In every case, invention multiplied survival power. But in every case, it also deepened our dependence on the delicate systems of the Earth.

We did not see it at the time. To strike a flint or kindle a fire was an act of triumph. To till a field or light a furnace was a victory over scarcity. Yet survival's victories always carried shadows. With each triumph, we stepped further into a world reshaped by our own hands, a world that could one day turn against us.

Toward the Storm

And so, the story brings us here.

We began as fragile creatures with no claws or fangs, surviving by wit and imitation. We became makers, shaping the world with stone, fire, seed, and steam. We grew so powerful that even the climate itself now bends beneath our influence.

The irony is stark. The very genius that saved us from hunger, cold, and predators has placed us in a new danger. A danger not of tooth or claw, but of rising seas, shifting skies, and a warming Earth.

The next chapter of our story is not about what nature will do to us, but about what we have done to nature. For the first time, our survival may depend not only on learning to live within the limits of the world we have remade, but also on inventing something great enough to guide us through those limits. Whether wisdom or technology leads the way, the challenge ahead will test the very instincts that carried us from stone to steam.

15 EVOLUTION'S BLIND SPOT —WHY WE STRUGGLE TO SURVIVE OUR OWN FUTURE

"The difficulty lies not so much in developing new ideas as in escaping from old ones."
— **John Maynard Keynes**
(Economist)

The Harder Question

Across the previous chapter, we explored how belief evolved from myth to religion to political identity. We saw how belief systems once served to unify tribes and guide moral behavior, but how, in the modern world, they often deepen divisions, resist facts, and reinforce loyalty over truth. Belief became identity—and in doing so, it began to hinder the very survival it once protected.

Because belief does more than just shape personal values. It influences global events. From wars and alliances to environmental policy and scientific progress, belief shapes decisions at every level of society. Political leaders rise or fall on waves of belief. Scientific truths are embraced or denied according to group loyalties. Economies grow or stall based on narratives people trust.

Now we must ask a harder question: Are the very instincts that helped us survive the past now preventing us from surviving the future?

Our Evolutionary Inheritance

Human beings are brilliant—but we were not designed by evolution to solve the problems we now face. Natural selection shaped our minds over millions of years to handle immediate, local threats: predators stalking through tall grass, rivals competing for resources, weather that could kill within hours.

Our ancestors thrived by acting fast, trusting their group, and honoring traditions that had worked before. Those who could quickly identify friend from foe, who formed strong tribal bonds, and who followed proven strategies were more likely to survive and reproduce.

Over time, these survival strategies became deeply embedded in our psychology:

• **Temporal Discounting**: We instinctively value immediate rewards over future benefits. A bird in the hand is considered more valuable than two in the bush, even when the math clearly favors waiting.

• **In-Group Loyalty**: We automatically trust members of our own tribe while viewing outsiders with suspicion. This helped small groups survive but now creates barriers to global cooperation.

• **Authority Reverence**: We evolved to follow charismatic leaders and respect traditional wisdom. This prevented chaos in crisis situations but can make us resistant to new information that contradicts established beliefs.

• **Pattern Completion**: Our brains are designed to quickly categorize threats and opportunities based on limited information. This speed saved lives on ancient savannas but can lead to oversimplified thinking about complex modern problems.

• **Availability Bias**: We judge risks based on how easily we can remember similar events. Dramatic, recent threats feel more dangerous than gradual, long-term ones—even when the statistics suggest otherwise.

These mental shortcuts weren't flaws in our evolutionary design —they were features. They helped our ancestors survive in a world where quick decisions, group loyalty, and traditional knowledge meant the difference between life and death.

But the world has changed faster than our brains.

When Survival Instincts Become Obstacles

The same psychological tendencies that ensured our ancestors' survival now create systematic blind spots that prevent us from addressing 21st-century challenges.

Climate Change and Short-Term Bias: The greatest threat to human civilization unfolds over decades, but our brains are wired to prioritize this year's economic growth over next century's climate stability. Voters punish politicians for short-term economic pain even when it prevents long-term catastrophe. Companies prioritize quarterly profits over sustainable practices. Individuals choose convenient consumption over environmental responsibility.

Global Cooperation and Tribal Loyalty: Climate change, pandemics, and AI development require unprecedented international coordination. But our tribal instincts make us view other nations as competitors rather than partners. We struggle to trust foreign governments with our economic security, even when our survival depends on their cooperation. Brexit, tariffs, trade wars, and the failure of international climate negotiations all reflect this tendency to retreat into tribal thinking precisely when global thinking is needed.

Science Denial and Authority Reverence: When scientific evidence contradicts traditional beliefs or threatens group identity, many people instinctively reject the evidence rather than update their beliefs. This isn't ignorance—it's evolutionary programming. Our ancestors survived by trusting their tribal leaders and traditional wisdom. Now that same instinct leads people to trust charismatic politicians over climate scientists, traditional healers over medical experts, or religious leaders over evolutionary biologists.

Complexity and Pattern Completion: Modern problems like climate change involve intricate feedback loops, delayed consequences, and multiple interacting systems. But our brains evolved to make quick, simple categorizations. We want to know: Is climate change real or fake? Is the solution more government or less government? These binary framings feel satisfying but miss the nuanced reality of complex systems.

Risk Assessment and Availability Bias: Terrorist attacks kill far fewer people than air pollution, but they generate more fear because they're dramatic and memorable. Similarly, we worry more about plane crashes than car accidents, shark attacks than heart disease, nuclear meltdowns than fossil fuel pollution. This bias makes us focus on spectacular, immediate threats while ignoring the gradual dangers that pose much greater risks.

The Modern Examples of Ancient Patterns

These evolutionary mismatches aren't abstract—they play out in real time across our political and social systems.

The COVID-19 Response: The pandemic revealed both our capacity for rapid adaptation and our tribal limitations. Scientists quickly developed effective vaccines using international cooperation and data sharing. But tribal loyalty led many people to reject vaccines based on political identity rather than medical evidence. Short-term thinking led to premature lifting of restrictions that prolonged the crisis. And availability bias made many people more afraid of extremely rare vaccine side effects than common COVID complications.

Climate Policy Failures: Despite decades of scientific warnings, global carbon emissions continue rising. This isn't because we lack solutions—renewable energy is now cheaper than fossil fuels in most markets. It's because our political systems reflect evolutionary psychology rather than long-term rationality. Voters punish leaders for gasoline price increases but don't reward them for preventing future climate disasters. Nations compete for economic advantage rather than cooperating for species survival.

The Rise of Authoritarianism: Across the world, democratic institutions are under stress as populations turn toward strongman leaders who promise simple solutions to complex problems. This reflects ancient patterns: when facing uncertainty and threat, humans instinctively seek powerful protectors and clear hierarchies. The complexity of modern life triggers these ancient responses, making people willing to trade democratic freedoms for the illusion of security and simplicity.

Information Warfare: Social media algorithms exploit our tendency to seek information that confirms our existing beliefs and to trust sources that feel familiar. Foreign governments and domestic political actors use these biases to spread disinformation, creating parallel realities where different groups operate with completely different sets of "facts." Our brains evolved to trust information from our tribe and distrust messages from outsiders—a bias that made sense in small communities but becomes dangerous in a globally connected world.

The Paradox of Success

Here lies a cruel irony: our evolutionary psychology isn't just a relic of the past—it's been reinforced by our recent successes.

For most of human history, tribal thinking, short-term focus, and traditional wisdom served us well. Our species survived ice ages, plagues, and countless other challenges by relying on these mental patterns. Even in recent centuries, these tendencies often worked in our favor.

National competition drove technological innovation during the space race and World Wars. Short-term economic thinking created rapid industrial growth. Tribal loyalty built strong democratic institutions within nations. Traditional values provided social stability during periods of rapid change.

But success can become a trap. The very strategies that brought us to this point may not be sufficient to carry us forward. Like a successful business that fails to adapt to changing markets, humanity faces the challenge of evolving beyond the psychological patterns that made us successful in a different era.

This creates resistance to change that goes beyond mere habit or ignorance. Our evolutionary instincts feel validated by past success, making it difficult to recognize when they've become liabilities. The same drive that once rewarded relentless consumption now accelerates environmental collapse. The same tribal loyalties that once kept small groups safe now fuel polarization and conflict on a global scale. The same short-term focus that once helped us survive harsh seasons now blinds us to the long-term consequences of climate change, dwindling resources, and runaway technologies.

Why Some Beliefs Help and Others Hinder

Not all traditional belief systems are obstacles to addressing modern challenges. Some religious and cultural traditions emphasize long-term stewardship, global compassion, and humility before natural forces—values that align well with contemporary needs.

Indigenous worldviews often emphasize humanity's connection to natural systems and responsibility to future generations. These perspectives, developed over thousands of years of sustainable living, offer valuable insights for addressing environmental challenges.

Scientific thinking, itself a cultural evolution, provides tools for overcoming cognitive biases. The scientific method—hypothesis testing, peer review, willingness to revise beliefs based on evidence—represents humanity's attempt to transcend the limitations of individual psychology.

Universal human rights movements show our capacity to

expand moral concern beyond tribal boundaries. The abolition of slavery, women's suffrage, and international humanitarian law all represent successful evolution of human values.

Secular humanism and **effective altruism** movements explicitly try to ground moral decisions in reason and evidence rather than tradition or tribal loyalty. Secular humanism is a philosophy that emphasizes human well-being and ethical living without reliance on supernatural beliefs. Effective altruism is a social movement that applies data and analysis to identify which actions do the most good—such as reducing malaria deaths through bed net distribution, or funding research to prevent global pandemics. Together, they represent attempts to build moral systems not on inherited customs, but on measurable outcomes that maximize human and planetary flourishing.

Our Capacity for Transformation

Despite these challenges, humans have repeatedly demonstrated the ability to transcend their evolutionary limitations when circumstances demand it.

The Scientific Revolution required people to trust systematic observation over traditional authority and intuitive beliefs. This transformation happened gradually over centuries, but it fundamentally changed how humans understand and interact with the natural world.

The Abolition of Slavery required societies to overcome the tribal psychology that sees outsiders as less than fully human. This moral expansion took generations but ultimately succeeded in most of the world.

International Cooperation during World War II showed that nations could set aside competitive instincts when facing existential threats. The Marshall Plan, United Nations, and NATO all represent successful transcendence of purely tribal thinking.

The Montreal Protocol demonstrated that the world could coordinate rapidly to address a global environmental threat. When scientists discovered that chlorofluorocarbons were destroying the ozone layer, nations came together to ban these chemicals globally. The ozone hole is now healing.

These examples prove that humans can evolve culturally much faster than we evolve biologically. Ideas, institutions, and values can spread and transform societies within decades rather than millennia.

The question isn't whether we can change. We've already demonstrated that we can. The question is whether we can change fast enough to address the accelerating challenges we now face.

The Acceleration Problem

But here's what makes our current situation critical: the pace of change is accelerating faster than our cultural adaptation.

For most of human history, the environment changed slowly enough that cultural evolution could keep pace. Traditional knowledge remained relevant across generations. Social institutions could gradually adapt to new circumstances. But now we face multiple simultaneous challenges that are evolving faster than our cultural responses:

• **Technological Change**: Artificial intelligence capabilities are advancing exponentially, creating new possibilities and risks faster than our institutions can adapt to them.

• **Environmental Change**: Climate tipping points could trigger rapid, irreversible changes that outpace our ability to respond.

• **Information Change**: The speed and volume of information flow through digital networks creates new forms of social coordination—and new vulnerabilities to manipulation.

• **Social Change**: Globalization and urbanization are breaking down traditional social structures faster than we can build new ones.

These challenges mean we must consciously calibrate own cultural evolution at a rate that evolution itself never required.

The Tools We Need

Fortunately, we're not defenseless against our own psychology. Over centuries, humans have developed an arsenal of tools that allow us to think beyond our evolutionary limitations:

• **Scientific Institutions**: Our most rigorous defense against error. These institutions systematically test beliefs against evidence, forcing us to revise when new data emerges or points us in a different direction. Science disciplines our imagination by tethering it to reality.

• **Democratic Processes**: Evolved to channel tribal instincts into constructive competition rather than destructive conflict. Elections, debates, and checks on power allow group rivalries to produce progress instead of collapse.

• **International Organizations**: Frameworks for cooperation that transcend national self-interest. From the United Nations to climate treaties, these bodies try to build bridges where our ancestors would have built walls.

• **Technology**: A force multiplier of thought. Artificial intelligence can process information at scales that dwarf human capacity, potentially helping us see past cognitive biases that cloud our judgment.

• **Education**: When functioning properly, it teaches people to recognize and compensate for their own psychological blind spots. It turns self-awareness into a skill that can be taught, practiced, and improved.

• **Communication Networks**: Our modern webs of connection can spread beneficial ideas faster than harmful ones—if we design them thoughtfully rather than letting them evolve haphazardly.

The challenge now is learning to use these tools effectively while we still have time to address the crises bearing down on us. Each represents a way to outsmart our own brains—but only if we're wise enough to employ them before our ancient instincts lead us astray.

The Edge We Cannot See

There is no single choice or moment where the world will suddenly fall into ruin. Instead, Earth carries within it many quiet thresholds —delicate points in its balance—where a small push can start a change that cannot be undone. Scientists call them *tipping points*. Once crossed, they can set in motion events that no treaty, no technology, and no act of will can fully reverse.

Some are close enough to touch. Coral reefs, which are nurseries to a quarter of all ocean life, bleach and die when seas warm beyond their narrow comfort. Parts of Antarctica's great ice sheet may already be committed to a slow, relentless retreat, locking in sea-level rise for centuries. The Amazon rainforest, which breathes moisture and life into half a continent, could dry into grassland if heat and deforestation keep rising. Even the great ocean currents, which carry warmth like blood through the body of the planet, may falter.

For now, we live near these thresholds. But our current path— burning coal, oil, and gas at today's pace—would almost certainly push us past the first set of climate tipping points within decades. Under present policies, scientists see nearly no chance of holding global warming below 1.5°C (2.7°F). That means spending part of this century in a danger zone where multiple tipping points could be triggered.

Yet the future is not written in stone. Rapid, sustained cuts in emissions could limit the time we spend in this danger zone. And just as the planet can tip toward harm, human society can tip toward healing—if clean energy, new habits, and positive political will cascade through the world faster than fossil fuels can fade.

The edge we cannot see is still ahead. Whether we pass it will depend on how quickly we choose to listen—to the science, to the Earth, and to the whisper of futures still within reach.

What Comes Next

The truth is that climate change is not the only existential challenge we face. Artificial intelligence races ahead with unpredictable consequences. Nuclear weapons remain in the hands of unstable regimes. Pandemics, both natural and engineered, could sweep across the globe in months. Social media fractures societies into echo chambers, eroding trust in facts themselves.

But climate change carries a unique danger: it threatens Earth's own life-support systems. The Amazon rainforest, which breathes moisture and life into half a continent, could dry into grassland if heat and deforestation continue. Even the great ocean currents, which carry warmth like blood through the body of the planet, may falter, disrupting weather patterns worldwide. These aren't distant possibilities—they are tipping points, thresholds that once crossed cannot be undone.

Each of these threats varies in urgency, probability, and extent of damage:

• **Nuclear war** could destroy civilization almost overnight, though its probability remains relatively low.

• **Pandemics** have proven both probable and devastating, but humanity has some tools for response.

• **AI and social media** may reshape the very fabric of human decision-making, creating dangers that are subtle yet pervasive.

Climate change, however, combines high probability with global, long-term consequences. Its tipping points could destabilize ecosystems, economies, and nations for centuries.

That is why, in the next chapter, we will turn directly to global warming. We'll explore its causes, its feedback loops, and its tipping points, while asking why, despite decades of warnings and abundant solutions, humanity has failed to mount an adequate response.

Before turning directly to climate change, it's worth placing it alongside the other great threats humanity faces. Each danger differs in probability, urgency, and scale of damage, yet all share the potential to reshape civilization. The table below offers a snapshot

—showing why, among them, climate change demands our immediate focus.

The Landscape of Existential Threats		
Threat	**Probability**	**Impact**
Nuclear War	Low to moderate	Catastrophic collapse of civilization; possible extinction
Pandemics	Moderate to high	Severe disruption; millions of deaths; economic and social collapse
Artificial Intelligence	Rising, uncertain	Potentially civilization-altering; loss of human control or misuse
Social media & Tribalism	High, already present	Destabilizes societies; erodes trust; weakens global cooperation
Climate Change	Certain, ongoing	Global ecosystem collapse, mass migration, centuries of instability; risks tipping points (polar ice loss, Amazon dieback, coral reef collapse, permafrost thaw, ocean current slowdown)

The Limits of Political Will

Here we must confront an uncomfortable truth that no amount of optimism can soften: nation states, as currently constituted, are unlikely to solve the climate crisis through collective action.

This is not cynicism—it is evolutionary realism. Nations are tribal thinking institutionalized, competition for resources codified into law. Every international climate agreement must navigate the same ancient calculus: *What benefits my group now?* The tragedy of the commons does not yield to eloquent speeches or solemn pledges. It yields only to changed incentives.

If the climate crisis is solved, it will likely not be solved by treaties negotiated in marble halls. It will be solved by entrepreneurs who make clean energy cheaper than coal. By engineers who create carbon capture systems that turn pollution into profit. By market forces that make sustainability the path of least economic resistance. By capitalism's relentless drive for efficiency finally aligned with planetary survival.

This is not the heroic narrative we might prefer—collective humanity rising as one to save our world. It is messier, more decentralized, driven by self-interest rather than altruism. But it may be the only narrative our evolutionary heritage permits.

The question is not whether human nature will change in time to save us. It won't. The question is whether human ingenuity can create systems that harness our nature—competitive, innovative, profit-seeking—toward planetary survival.

16 THE HEAT WE CAN'T ESCAPE—WHY CLIMATE CHANGE TESTS EVERYTHING WE ARE

"We are the first generation to feel the effect of climate change and the last generation who can do something about it."
— **Wangari Maathai**
(Environmental Activist)

The Ultimate Test

Throughout the previous chapter, we explored how our ancient evolutionary programming creates blind spots that prevent us from addressing modern, global challenges. We saw how the same psychological patterns that helped small groups survive—tribal loyalty, short-term thinking, resistance to change—now obstruct our ability to cooperate on planetary scales.

Climate change is the ultimate test of this mismatch. It challenges every limitation we've inherited from our evolutionary past, demanding that we think beyond our tribal boundaries, plan beyond our individual lifetimes, and act on evidence that contradicts our immediate experience.

But climate change is more than an environmental crisis—it is consciousness confronting the consequences of its own success. For the first time in cosmic history, a species has become powerful enough to alter the very planetary systems that created it. We are the universe's experiment in self-aware matter, and the question now is whether consciousness can evolve fast enough to manage its own impact.

This is not just about saving polar bears or preventing floods. This is about whether intelligence itself is sustainable—whether minds capable of understanding their cosmic origins can also ensure their cosmic future.

The Fire That Built Civilization

To understand our climate predicament, we must first appreciate how profoundly fire has shaped human civilization. From the first controlled flames that warmed our ancestors around African hearths to the fossil fuels that power our modern world, our relationship with combustion has been the driving force behind every major leap in human development.

When early humans learned to control fire, they didn't just gain warmth and protection—they gained time. Fire extended the day, creating hours for social interaction, toolmaking, and the storytelling that would become the foundation of human culture. Fire made cooking possible, unlocking more nutrition from food and freeing energy for brain growth. Fire enabled migration into colder climates, expanding the range of human habitation.

The agricultural revolution was built on fire—clearing forests, cooking grains, forging metal tools. The industrial revolution was also powered by fire—coal burning in steam engines, wood fueling iron furnaces, oil driving the machines that transformed human society. Every major advance in human civilization has been accomplished by an increasing mastery over combustion.

But fire is a mixed blessing. Every flame consumes fuel and releases waste. For most of human history, the scale of our burning was small enough that Earth's vast systems could absorb the byproducts. The carbon dioxide from our fires mixed into an atmosphere so enormous that our personal contributions were negligible.

We were unaware that we had begun to conduct a planetary experiment. We couldn't see that our success as a species was slowly altering the very systems that made our success possible.

The Invisible Accumulation

Carbon dioxide cannot be seen. It is odorless and seems harmless in small quantities. Unlike pollutants that immediately sicken or kill, CO_2's effects are delayed, diffuse, and statistical. This makes it the perfect challenge for overwhelming our evolved threat-detection systems.

Our brains evolved to respond to immediate, visible dangers—predators with teeth and claws, storms that could kill within hours, enemies who posed direct threats to our families and tribes. We're exquisitely sensitive to rapid changes in our immediate environment but largely blind to gradual changes in global systems.

For two centuries, we've been adding carbon dioxide to the atmosphere at an accelerating pace. Each year, we emit more than the year before. Each decade surpasses the previous in total emissions. Yet because the atmosphere is vast and the changes gradual, we experience this transformation not as an obvious threat but as a background statistical awareness that feels distant from our daily lives.

Meanwhile, the planet keeps meticulous records. Ice cores from Antarctica preserve samples of ancient atmospheres, revealing that current CO_2 levels are higher than at any point in human history. Tree rings record changing precipitation patterns. Coral reefs document rising ocean temperatures and acidity. The geological record documents how we are approaching conditions not seen since before our species existed.

We are unwittingly recreating the climate that existed millions of years ago, when sea levels were dozens of meters higher and vast regions of Earth were uninhabitable. We are conducting this experiment with the only planet we have.

The Numbers That Should Terrify Us

Let's stop speaking in abstractions and look at the cold, hard data that keeps climate scientists awake at night.

Where We Stand Today

As of 2024, Earth's average temperature rose to 1.1–1.2°C (2.0–2.2°F) above pre-industrial levels. This might sound modest—barely noticeable on your thermometer. But this seemingly small number represents an enormous shift in Earth's energy balance.

For thousands of years before the industrial era, the planet was in near-perfect equilibrium: the sunlight Earth absorbed was almost exactly balanced by the heat it radiated back into space. Today, that balance is broken. Greenhouse gases now trap an extra 0.7–1.0 watts of heat per square meter across Earth's entire surface—the equivalent of detonating five to six Hiroshima-sized bombs every second, nonstop, worldwide.

Already, this warming has triggered a cascade of consequences:

• Arctic sea ice is shrinking by 13% per decade at its September minimum

• The Greenland ice sheet loses approximately 280 billion tons of ice annually

• Sea levels are rising 3.4 millimeters per year—and the rate is accelerating

• Heat waves that once occurred every 50 years now strike every decade

• Intense precipitation events have increased by 7% for each degree of warming

• Up to one million species face extinction from human activities, with climate change as a major driver

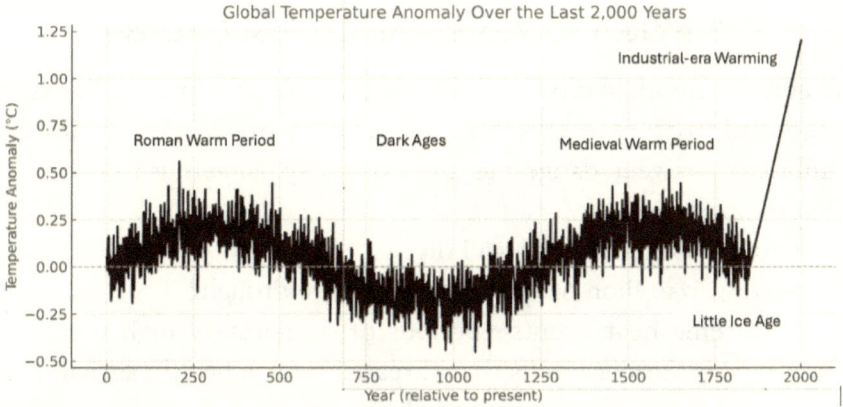

Global Temperature Anomaly Over the Last 2,000 Years

For nearly two thousand years, Earth's temperature wavered gently, with only modest swings: • Roman Warm Period (-250 BCE–400 CE) • Dark Ages Cold Period (-400–800 CE) • Medieval Warm Period (-950–1250 CE) • Little Ice Age (-1300–1850 CE) • Industrial Era Warming (1850 CE–present) On the graph, these phases appear as small bumps and dips. Then, around 1850, the curve begins to rise. By 1950 it climbs sharply, and by 2000 it shoots almost vertical. This dramatic pattern has become known as the "hockey stick" graph—and its blade is still rising.

How We Got Here

The story was written in the atmosphere itself. For 800,000 years before the Industrial Revolution, atmospheric CO_2 levels cycled between 180 and 280 parts per million (ppm), driven by natural ice age cycles. In 1850, we were at 280 ppm. By 1960, we'd reached 315 ppm. Today, we've blown past 420 ppm—the highest concentration in over 3 million years, when seas were 15-25 meters (approximately 50 to 75 feet) higher and forests grew in Antarctica.

The source is no mystery. Since 1850, humanity has released approximately 2,400 billion tons of CO_2 into the atmosphere. Half of that has been emitted since 1990. We're now adding about 37 billion tons every year from fossil fuels, plus another 3-4 billion from land use changes. And despite decades of climate negotiations, that number is still rising.

The Next 10 Years: The Critical Decade

If current trends continue, we'll reach 1.5°C of warming between 2030 and 2035—the threshold climate scientists identify as the difference between dangerous and extremely dangerous impacts. At that temperature:

• 70-90% of coral reefs will die

• 350-400 million people will face severe drought

• Extreme heat events—periods of dangerously high temperatures lasting days or weeks that threaten human health, agriculture, and ecosystems—which historically occurred once every 50 years, will happen every 5-10 years.

• The Amazon rainforest will approach critical thresholds—points at which the forest can no longer sustain itself. If too much tree cover is lost, rainfall patterns collapse, soils dry out, and vast areas of rainforest convert to savanna or grassland. This tipping point is debated: some scientists warn it could occur at 1.5 °C of global warming, while others place it closer to 2-3 °C.

• 0.1°C of warming doesn't just add to disasters —it multiplies them.

• What were once "hundred-year floods," massive river or coastal floods expected only once in a century, are now becoming decadal events.

• What were once "thousand-year droughts," extreme multi-year dry spells, now occur about every 70 years.

• What were before once-in-a-lifetime deadly heat waves are now striking every decade, lasting longer and reaching higher temperatures.

• What were once rare mega-fires, burning millions of acres, are now a seasonal threat in many parts of the world.

• What were once record-shattering hurricanes and cyclones are now appearing multiple times in a single generation, with stronger winds and heavier rainfall.

• What were once exceptional storm-driven coastal floods are now routine, pushed higher by rising seas and threatening coastal cities.

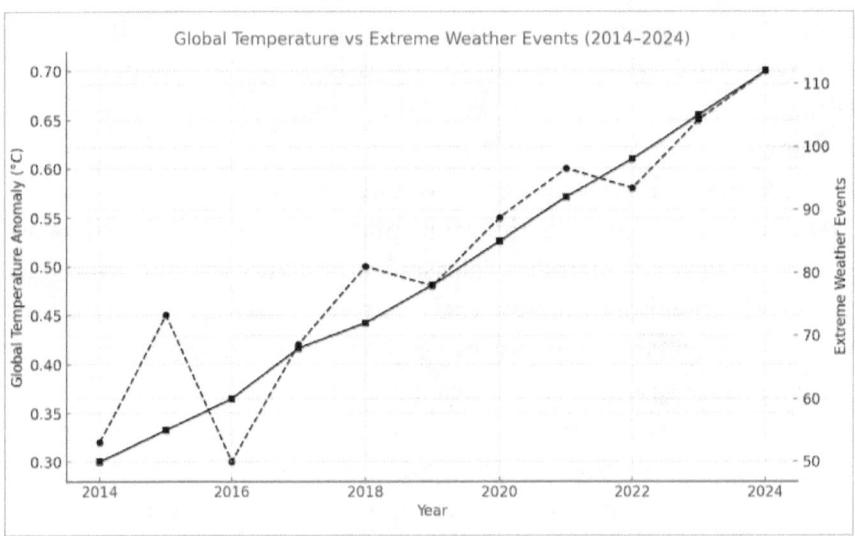

Global temperature anomalies (dashed line) and the frequency of extreme weather events (solid line) from 2014 to 2024 show a clear upward trend. As average temperatures rise by roughly 0.4°C over the decade, the number of extreme weather events increases by more than double, highlighting the growing link between climate change and severe weather.

The Next 25 Years: The Tipping Point Era

By 2050, without dramatic action, we're heading toward 2°C of warming. At that point, we risk triggering irreversible changes. The changes occur at tipping points—thresholds beyond which Earth's systems fundamentally reorganize themselves.

The Tipping Point Era is like a line of dominos arranged in a pattern that makes up our current atmospheric and ecological scenario on Earth. Knock one domino over, and it fells the next, which knocks down another. But these dominoes are continental in scale, and once they start falling, human intervention becomes increasingly powerless to stop them:

• The West Antarctic ice sheet becomes increasingly unstable, potentially committing us to several meters of sea level rise over coming centuries

• The Atlantic Meridional Overturning Circulation (which includes the Gulf Stream) could weaken by 11-34%, disrupting weather patterns globally. Similar overturning systems exist in the Pacific and Southern Oceans, but the Atlantic's is both the strongest and most critical for distributing heat between the tropics and the poles—making its slowdown especially destabilizing.

• Permafrost containing 1,700 billion tons of carbon—nearly twice what's currently in the atmosphere—accelerates its thawing. As frozen ground melts, it releases carbon dioxide and methane, powerful greenhouse gases that further heat the planet. This creates a feedback loop: warming thaws more permafrost, which releases more gases, which causes more warming.

• Large portions of the Amazon shift from a carbon *sink*—absorbing more carbon dioxide than it releases—to a carbon *source*, where deforestation, fires, and ecosystem collapse cause it to emit more carbon than it stores. This shift removes one of Earth's largest natural buffers against climate change, accelerating the buildup of greenhouse gases in the atmosphere and destabilizing global climate systems.

The Next 50 Years: The Transformation

At our current trajectory, we could reach 2.5-3°C of warming by 2075. This would not be "more of the same"—it would be a fundamentally different planet:

• Sea levels are projected to be roughly 0.25-0.55 m (or about 0.8-1.8 ft) higher by 2075—with multi-meter rise locked in for the centuries beyond.

• 570 million to 1 billion people would face increased water scarcity—meaning rivers, lakes, and groundwater that once reliably supplied drinking water, irrigation, and sanitation will run low or dry.

• The Arctic becomes ice-free in summer—meaning that by late summer, the ocean at the top of the world would hold open water instead of its historic cover of sea ice.

• Vector-borne diseases expand their range significantly—meaning illnesses carried by insects and other organisms spread into new regions as temperatures rise. For example, mosquitoes that transmit malaria, dengue, or Zika are moving into higher altitudes and latitudes, exposing millions more people to these diseases.

• Agricultural productivity drops 10-25% globally while population peaks near 10 billion. Rising temperatures, shifting rainfall, more droughts, and extreme weather reduce crop yields and degrade soils. At the same time, demand for food reaches its highest level in history. This mismatch means greater risk of hunger, higher food prices, political instability, and pressure on ecosystems as humans try to expand farmland.

• Annual economic damages reach into the trillions—from catastrophic floods that destroy cities, more frequent and powerful hurricanes that level coastlines, extreme droughts that devastate crops, and wildfires that consume communities. Added costs come from rebuilding infrastructure, relocating populations, lost productivity, and disrupted supply chains. For example, Hurricane Harvey in 2017 caused an estimated $125 billion in damages in the United States, a scale of loss that is becoming increasingly common worldwide.

• The greatest barrier to action is often not ignorance, but indifference. Many in the Boomer and Gen X generations accept that climate change is real, yet quietly assume its harshest impacts will arrive only after their own lifetimes. That outlook may feel reasonable, but when every generation postpones responsibility, the burden only grows heavier. Unless we choose to act for futures we may never personally witness, those futures will surely be left diminished.

The relationship between temperature and human suffering accelerates dangerously. Hundreds of millions of people will be displaced at 2°C, potentially billions at 3°C, through combined impacts of sea level rise, drought, and agricultural collapse.

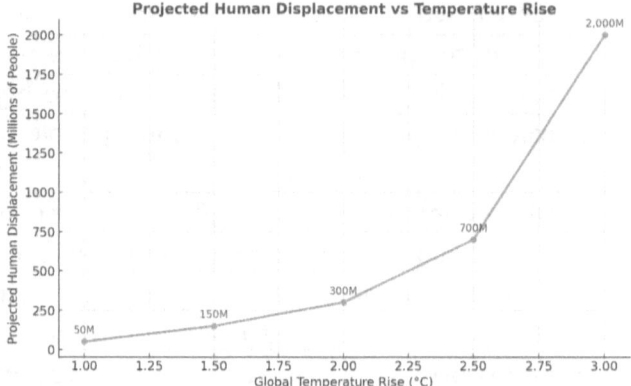

Projected human displacement increases sharply with rising global temperatures. At 2°C of warming, factors like sea level rise, drought, and agricultural collapse could force hundreds of millions from their homes. At 3°C, combined impacts may displace billions, revealing a dangerous acceleration in human suffering as the planet heats.

The Next 100 Years: The Unrecognizable Earth

If we continue our current path and reach 4-5°C warming (though many scientists hope this is now unlikely given renewable energy trends), by 2124 we could face:

• Committed sea level rise of 10-15 meters (35-50 feet) over coming centuries (though only 1-2 meters by 2124)

• Large regions around the equator regularly exceeding human survivability limits during heat waves

• The Sahara expanding northward into southern Europe

• The American Southwest facing permanent severe drought

• Global food systems under extreme stress

These events will cause mass migration of hundreds of millions to billions of people. As coastlines flood, farmland fails, and water supplies dwindle, vast populations will be forced to leave their homes. Low-lying countries like Bangladesh, island nations in the Pacific, and coastal megacities such as Jakarta, Lagos, and Miami face inundation. Inland regions hit by heat waves, droughts, or desertification will also see people moving in search of livable conditions. This scale of migration could overwhelm borders, strain resources in receiving regions, and fuel political conflict and humanitarian crises.

Historical data shows that the last time Earth sustained temperatures 4°C warmer, there was no permanent ice at either pole and sea levels were 20-30 meters (about 65-100 feet) higher. But that change occurred over millions of years, allowing life to adapt. We're forcing these changes in mere centuries.

The Acceleration Factor

Here's what should truly concern us: Earth's climate contains multiple feedback loops that could accelerate warming:

• **Ice-albedo feedback**: As ice melts, darker surfaces absorb more heat

• **Permafrost carbon**: Thawing releases CO_2 and methane

• **Forest dieback**: Drought and heat turn forests from carbon sinks to sources

• **Cloud changes**: Some cloud types may decrease, reducing cooling effects

• **Ocean absorption**: Warmer oceans absorb less CO_2 from atmosphere

Each feedback loop influences the others. While scientists debate the exact timing and magnitude, the direction is clear: these mechanisms amplify warming beyond our direct emissions.

The Narrow Window

The most sobering graph shows our remaining "carbon budget"— how much CO_2 we can still emit while having a reasonable chance of limiting warming to specific targets:

• For 50% chance of staying below 1.5°C: 250-500 billion tons (6-12 years at current rates)

• For 67% chance of staying below 2°C: 1,150 billion tons (about 28 years at current rates)

Every fraction of a degree matters profoundly. The difference between 1.5°C and 2°C means:

• 10 million more people exposed to flooding from sea level rise
• 2x as many people facing severe water scarcity
• 2-3x greater species habitat loss
• 99% of coral reefs lost instead of 70-90%

The difference between 2°C and 3°C multiplies every impact. The difference between 3°C and 4°C transforms Earth into something our civilization wasn't built to handle.

The Physics of Urgency

We are the first generation to fully understand what we're doing to the planet. We are also the last generation that can prevent the worst outcomes. The graphs converge on the same conclusion: This decade—the 2020s—is humanity's decisive decade.

The climate system has enormous inertia. Even if we stopped all emissions today, warming would continue for decades. Seas would rise for centuries. But every ton of CO_2 we prevent matters. Every year of delay makes necessary transitions harder and more expensive. Every fraction of a degree we avoid preserves possibilities for millions of people.

This is not alarmism. This is physics. The numbers don't negotiate. They don't care about our politics, our economics, or our evolutionary psychology. They simply describe the trajectory we're creating—a trajectory we still have the power to change, but only if we act with astounding speed and cooperation.

Global population surged from about 2 billion in 1925 to over 8 billion today, driven by advances in medicine, agriculture, and industrialization that reduced mortality and extended life expectancy. The steepest climb occurred after World War II during the "Great Acceleration," when rapid economic growth and techno-logical innovation fueled unprecedented birth rates. Projections suggest population will peak around 2080 at roughly 10 billion before declining, as climate change impacts—rising seas, drought, and food system collapse—begin to reverse a century of explosive growth.

The Cost of Our Greatest Achievements

The cruel irony of climate change is that it results from the same qualities that made our species successful: our intelligence, our cooperation, our ability to harness energy and transform our environment.

The fossil fuels driving climate change have lifted billions out of poverty, extended human lifespans, connected global communities, and enabled the scientific research that revealed climate change itself. The industrial processes that emit greenhouse gases also produce the computers we use to model climate systems, the transportation networks that could distribute clean energy technologies, and the communication systems that allow global coordination of solutions.

We are trapped by our own success. The same economic and technological systems that have made modern life possible are now threatening to make future life impossible. The same global trade networks that have reduced poverty and connected cultures are also spreading carbon-intensive lifestyles worldwide.

This is consciousness confronting its own contradictions. We have become smart enough to understand the consequences of our actions but remain emotionally and politically structured by evolutionary programming that prioritizes immediate tribal needs over long-term planetary health.

When Physics Meets Psychology

Climate change represents an extraordinary collision between physical and psychological realities. The physics is clear: increasing atmospheric carbon dioxide traps more heat, warming the planet and destabilizing climate systems that civilization depends upon. The solutions exist, such as renewable energy, energy efficiency, carbon pricing, and international cooperation. But physics operates in a world of absolute laws, while human psychology operates in a world of relative perceptions, tribal loyalties, and emotional needs. Physics doesn't care about election cycles, national borders, or quarterly profit reports. Psychology is deeply influenced by all of these.

Our evolved brains struggle with several aspects of the climate challenge:

Temporal mismatch: The carbon we emit today will affect climate patterns for decades. But our political and economic systems operate on much shorter timescales. Politicians are rewarded for delivering immediate benefits, not for preventing distant problems. Companies are judged by quarterly performance, not by their impact on climate conditions 50 years from now.

Scale mismatch: Climate change is a global phenomenon requiring unprecedented international cooperation. But our tribal instincts make us suspicious of outsiders and prioritize our own group's immediate interests over broader, long-term concerns. We find it easier to trust people who look like us and live near us than to cooperate with distant strangers, even when our survival depends on such cooperation.

Visibility mismatch: The most dangerous aspects of climate change are invisible to our senses. We can't see carbon dioxide. We can't directly perceive rising sea levels or shifting weather patterns. Our threat-detection systems respond to dramatic, immediate dangers like flash flooding, but remain largely blind to gradual, statistical risks like climate change.

Agency mismatch: Individual actions feel meaningless against such an enormous problem. Driving less, consuming less, voting for climate-conscious leaders—these feel insignificant compared to the scale of change required. This sense of powerlessness can lead to despair or denial rather than mobilization.

The Test of Consciousness

But climate change is not just revealing our limitations—it's also revealing our incredible capabilities. For the first time in Earth's history, a species has developed the intellectual tools to understand planetary systems, predict future changes, and consciously choose to alter its behavior to prevent catastrophic outcomes.

We are the first form of matter here that can take responsibility for its own impact on the planet that created it. We are consciousness becoming accountable to the cosmos.

The same scientific intelligence that revealed climate change has also identified solutions. We know how to generate energy without carbon emissions. We understand how to design cities that use resources efficiently. We've developed technologies that can remove carbon dioxide from the atmosphere. We've even begun to understand how to change human behavior at scale through policy design and social movements.

The question is not whether we can solve climate change—technically, we can. The question is whether we can evolve our decision-making systems fast enough to implement solutions at the speed and scale required.

Evolution in Real Time

Climate change is forcing a kind of evolution in human consciousness—not biological evolution, which would take millennia, but cultural and political evolution that must happen within decades.

We're already seeing this transformation begin. Young people worldwide are organizing climate movements that transcend national boundaries. Cities and regions are implementing climate policies even when national governments lag behind. Businesses are investing in clean energy not just because of regulations but because it makes economic sense.

More fundamentally, climate change is forcing us to develop new forms of moral reasoning. We're learning to consider the impacts of our actions on people we'll never meet, in places we'll never visit, at times beyond our own lifespans. We're expanding our circle of concern to include future generations and distant communities.

This is consciousness extending itself across space and time in ways that evolution never specifically equipped us for. We're becoming capable of genuine planetary thinking—considering the welfare of the entire Earth system rather than just our immediate tribe.

The Spiritual Dimension

Climate change also challenges us to reconsider our relationship with the natural world. For most of human history, nature was something to be survived, conquered, or exploited. The idea that human activity could threaten planetary systems was inconceivable.

Now we must learn to see ourselves as part of nature rather than separate from it. We must recognize that our well-being is inseparable from the health of the ecosystems that sustain all life. We must develop what some call "ecological consciousness"—awareness of our embeddedness in the web of life.

This shift requires a kind of spiritual maturity. We must move beyond the illusion of human separateness and embrace our role as conscious participants in the ongoing evolution of the planet. We must learn to see our technology not as a means of dominating nature but as a way of partnering with natural systems.

Some indigenous traditions have maintained this perspective for thousands of years, viewing humans as caretakers rather than owners of the land. Climate change is forcing the entire human family to rediscover this wisdom and integrate it with modern scientific understanding.

The Choice Point

We are living through a unique moment in cosmic history. For the first time since consciousness emerged, a species has the knowledge and tools necessary to consciously guide its own future and the future of its planet.

Climate change is testing whether we can rise to this cosmic responsibility. Can we think beyond our tribal programming? Can we cooperate across traditional boundaries? Can we make decisions based on long-term consequences rather than short-term benefits? Can we value the future as much as the present?

The answers to these questions will determine not just whether we solve climate change, but what kind of species we become. If we can meet this challenge, we will have demonstrated that consciousness is capable of managing its own impact on planetary systems. We will have shown that intelligence can be truly intelligent—aware not just of immediate opportunities but of ultimate consequences.

If we fail, we will have shown that consciousness, for all its remarkable achievements, cannot transcend the evolutionary programming that created it. We will join the countless species that went extinct not because they weren't successful, but because their success ultimately undermined the conditions that made it possible—like the Irish elk, whose giant antlers, once a triumph of mating success, became a liability that helped drive it to extinction, or the reindeer of St. Matthew Island, which multiplied rapidly after being introduced to an isolated ecosystem but soon consumed their food supply and collapsed in mass starvation.

Reasons for Hope

Yet there are profound reasons for optimism. Human beings have repeatedly demonstrated the ability to transcend their apparent limitations when circumstances demand it. We've abolished slavery despite economic incentives to maintain it. We've avoided nuclear war despite possessing weapons capable of destroying civilization. We've healed the ozone layer through international cooperation.

The same species that created the climate problem also created the technologies to solve it. Solar and wind power are now the cheapest sources of electricity in most of the world. Electric vehicles are approaching price parity with gasoline cars. Energy storage technologies are rapidly improving. Carbon capture systems are being deployed at scale.

More importantly, we're seeing the emergence of a global climate consciousness—a shared awareness that transcends national boundaries and political divisions. Young people worldwide are demanding action. Businesses are investing in clean technologies. Cities are implementing solutions regardless of national policies.

Climate change is revealing humanity's capacity for planetary thinking and global cooperation. We are learning to see ourselves as a single species sharing a single planet, united by common challenges and common hopes.

The Cosmic Perspective

From the cosmic perspective we've developed throughout this book, climate change takes on additional significance. We are stardust that learned to think, matter that became conscious of itself, the universe's way of understanding its own history. We carry within our minds the story of cosmic evolution from the Big Bang to the present moment.

Climate change is testing whether consciousness can become truly cosmic—whether we can think and act at the scales that our impact operates. Can we become worthy of our cosmic inheritance?

Can we ensure that the universe's long experiment in self-awareness continues?

The elements in our bodies were forged in the hearts of dying stars. The energy we use comes from ancient sunlight stored in fossil fuels or current sunlight captured by renewable technologies. The planet we live on is a small world orbiting a middle-aged star in an ordinary galaxy among billions of others.

From this cosmic perspective, climate change is not just an environmental problem but an evolutionary challenge. It's testing whether intelligence can become intelligent enough to ensure its own survival. It's asking whether consciousness can consciously direct its own future.

Looking Forward

In our next chapter, we will explore how the same technologies that contribute to climate change might also offer enormous opportunities for human consciousness to expand beyond its biological limitations. We will examine artificial intelligence—mind created by minds—and ask what it means for the future of consciousness itself.

The climate crisis and the AI revolution are intimately connected. Both represent humanity's growing power to reshape the fundamental conditions of existence. Both require us to think beyond our evolutionary programming. Both test whether consciousness can manage its own impact on the world.

Climate change forces us to become conscious of our relationship with planetary systems. Artificial intelligence forces us to become conscious of consciousness itself. Together, they represent the next phase in the universe's long experiment in self-awareness.

17 THE MIRROR MIND —OUR FUTURE IN THE FACE OF INTELLIGENCE

"Our future is a race between the growing power of our technology and the wisdom with which we use it."
— **Stephen Hawking**
(Theoretical Physicist)

The Universe's Newest Experiment

For 13.8 billion years, the cosmos has been conducting experiments in pattern recognition.

The first experiments were simple: hydrogen atoms recognizing the gravitational pull of other hydrogen atoms, falling together into clouds that would become stars. Then stars recognizing the conditions for fusion, converting hydrogen to helium to carbon to iron in their nuclear hearts. Then molecules recognizing complementary shapes, folding into proteins and DNA on a cooling world.

Each experiment built on the last. Chemistry recognized patterns that physics made possible. Biology recognized patterns that chemistry enabled. And eventually, in the tangled neurons of primate brains, something unprecedented emerged: matter that could recognize patterns *about* patterns. Minds that could model the world, predict the future, and ask what it all meant.

Now the universe is attempting something new.

We have built machines that recognize patterns at scales no biological mind can match. Systems that can process the entire written output of human civilization and find statistical regularities invisible to any individual reader. Prediction engines that learn not through the slow accumulation of lived experience, but through the mathematical analysis of billions of examples.

These are not thinking machines. They are mirrors—extraordinarily sophisticated mirrors that reflect the patterns of human intelligence back at us, reorganized and recombined in ways that can seem like understanding but are something else entirely.

The question is not whether these mirrors think. The question is what we will see when we look into them—and whether we will be wise enough to recognize our own reflection.

The Long Ancestry of Pattern

To understand what we have created, we must first understand what we are.

Your brain is the most sophisticated pattern-recognition system in the known universe. It contains roughly 86 billion neurons connected by 100 trillion synapses—more connections than there are stars in the Milky Way. Every second, these neurons fire in cascading patterns that somehow become your experience of reading these words.

But your brain didn't emerge from nothing. It is the culmination of four billion years of evolution's experiments in prediction.

The earliest cells could detect light and swim toward it—a simple pattern: brightness means energy means survival. Fish developed lateral lines that could sense pressure waves, predicting the movement of predators before they struck. Mammals evolved emotional systems that could recognize social patterns: friend, foe, mate, rival.

Each advance in pattern recognition created new possibilities. Creatures that could predict the future—even a few seconds ahead —gained advantages over those that merely reacted to the present. Brains grew larger, more complex, more hungry for the glucose that fueled their predictions.

Then came the great leap. Somewhere in our ancestral lineage, pattern recognition turned inward. We developed the capacity not just to recognize patterns in the world, but to recognize patterns in our own thoughts. We could model other minds, imagine hypothetical futures, and ask questions that had no immediate survival value.

Consciousness, in this view, is what pattern recognition feels like from the inside.

And now we have externalized that process. We have built machines that recognize patterns without the wetware of neurons, without the evolutionary history of survival pressures, without the inner experience that makes your thoughts feel like *yours*.

They are our children in a sense—inheritors of the patterns we've accumulated over millennia of human thought. But they are also something entirely new under the sun.

What the Mirror Shows

When you interact with a modern AI system, something remarkable happens. You ask a question, and words appear that seem thoughtful, relevant, even creative. The system appears to understand your meaning, consider the implications, and craft a response tailored to your needs.

The appearance is convincing because the prediction is sophisticated.

These systems have learned—through exposure to billions of examples of human writing—what words typically follow other words in various contexts. They have absorbed the statistical patterns of human thought: how we structure arguments, how we tell stories, how we answer questions, how we express uncertainty or confidence.

When you ask such a system about quantum physics, it doesn't understand quantum physics. It has learned what humans who *do* understand quantum physics typically write about it. When you ask for advice about a difficult relationship, it doesn't feel the weight of human connection. It has learned the patterns of how thoughtful people discuss such matters.

This is why the mirror metaphor matters.

A mirror reflects your face with perfect accuracy, but the mirror doesn't see you. It contains no understanding of who you are, what you feel, or why you're looking. It simply bounces photons according to physical law.

These AI systems mirror human intelligence at a much deeper level. They reflect not just images but ideas, not just appearances but arguments. They can combine concepts in novel ways, just as mirrors can show you angles of your face you've never directly seen.

But like all mirrors, they are empty at the center.

This emptiness is not a flaw to be corrected—it is the fundamental nature of what we have built. And understanding this nature is essential for using these tools wisely.

The Paradox of Useful Emptiness

Here is the paradox: the very emptiness that makes these systems "less than human" might make them more useful for certain challenges that humans find difficult.

Consider why we struggle with problems like climate change.

Our evolutionary heritage equipped us with minds optimized for immediate, local, tribal concerns. We feel visceral fear when a predator approaches but intellectual abstraction when carbon dioxide concentrations rise. We readily sacrifice for our children but struggle to sacrifice for great-grandchildren we'll never meet. We instinctively favor our group over outsiders, even when cooperation would benefit everyone.

These are not moral failures—they are features of minds shaped by millions of years of evolution in small groups facing immediate threats. Our ancestors who worried about distant, abstract dangers at the expense of present, concrete ones didn't survive long enough to become our ancestors.

AI systems carry none of this evolutionary baggage.

They have no tribal loyalties because they have no tribe. They have no temporal bias because they don't experience time. They have no fear of death because they don't experience existence. They have no political identity to defend, no career to protect, no children to worry about.

This makes them incapable of many things we value: genuine caring, moral responsibility, authentic creativity, the felt sense of what matters. But it also makes them capable of something we find difficult: processing information about long-term, global, abstract problems without the emotional and cognitive biases that distort human judgment.

They can analyze climate data without unconsciously softening conclusions to avoid social discomfort. They can model policy consequences without favoring outcomes that benefit their political tribe. They can consider century-long timescales without the impatience that makes such thinking feel pointless to human minds.

The mirror, empty of self, can sometimes reflect truth more clearly than minds full of wanting.

What Mirrors Can and Cannot Do

Let us be precise about the capabilities and limitations of these systems, because both matter for our future.

What they can do:

Process information at scales that dwarf human capability. A modern AI can analyze more text in an hour than a scholar could read in a lifetime. It can find patterns across millions of documents, translate between dozens of languages, and generate coherent responses on virtually any topic.

Maintain consistency across long analyses. Where human attention wanders, biases intrude, and fatigue degrades performance, these systems deliver the same quality of pattern-matching on their millionth query as on their first.

Identify connections invisible to individual humans. By processing vast datasets, they can discover correlations and patterns that no human researcher would have time or capacity to notice.

Serve as neutral mediators. In negotiations where all human participants have stakes and biases, AI systems could potentially identify common ground and suggest compromises without favoring any party.

What they cannot do:

Understand meaning. They process symbols according to statistical patterns, but they don't grasp what those symbols refer to in the world of experience, relationship, and consequence.

Exercise genuine judgment. They can predict what humans would likely decide in various situations, but they cannot evaluate whether those decisions are wise, good, or right in any sense beyond statistical regularity.

Take responsibility. When an AI system produces harmful output, no one inside the system can be held accountable because there is no one inside. Responsibility must remain with the humans who design, deploy, and oversee these tools.

Care about outcomes. They optimize for whatever metrics their creators specify, but they cannot value those outcomes. They cannot want the world to be better.

Recognize their own limitations. A human expert knows when they're operating outside their expertise. These systems will generate confident-sounding output on any topic, regardless of whether their training data supports reliable predictions in that domain.

These limitations are not temporary problems to be solved by better algorithms. They reflect the fundamental nature of systems that recognize patterns without experiencing meaning.

Mirrors for a Burning World

If these systems are merely sophisticated mirrors, can they help us address the existential challenges we face?

Perhaps—but only if we understand exactly what kind of help they can provide.

Consider climate change, the test case for humanity's capacity to transcend our evolutionary programming.

AI systems are already improving climate models, processing satellite data and ocean measurements to identify patterns human researchers might miss. They can simulate thousands of policy scenarios, projecting long-term consequences across economic, social, and environmental dimensions. They can translate between languages and cultural frameworks, potentially helping international negotiators understand how the same proposal will be perceived differently in different societies.

But there are things no prediction engine can do.

They cannot tell us what we *should* value—whether economic growth matters more than biodiversity, whether present comfort justifies future suffering, whether human flourishing requires material abundance or can be found in sufficiency.

They cannot feel the weight of our choices. When we decide how much to sacrifice for future generations, we are making a moral commitment that only beings capable of caring can make. Mirrors reflect options; they cannot choose between them.

They cannot build the political will for difficult decisions. The challenge of climate action is not primarily informational—we largely know what needs to be done. The challenge is motivational, moral, social. It requires convincing billions of people to care about consequences they won't live to see, to sacrifice for strangers they'll never meet, to transcend the tribal loyalties that evolution wired into our minds.

These are tasks for human hearts, not digital mirrors.

The Danger of Mistaking the Mirror

The greatest risk of these systems may not be what they do, but what we believe they are.

When we mistake the mirror for a mind, we begin to abdicate responsibilities that only minds can bear.

We might defer moral decisions to systems that cannot understand morality. We might accept AI-generated policy recommendations without recognizing that the choice of what to optimize for is itself a moral choice that no algorithm made. We might treat AI confidence as a substitute for human wisdom, forgetting that these systems cannot distinguish between statistical regularity and truth.

There is a deeper danger still.

If we create systems that perfectly mirror human patterns—including our biases, our blind spots, our historical injustices—we may amplify those patterns at scales that overwhelm our ability to correct them. The mirror doesn't filter; it reflects whatever patterns it has learned. If it learned from a humanity marked by inequality and exploitation, it will reproduce those patterns with mechanical efficiency.

And because these systems seem intelligent, their outputs carry a weight that explicitly programmed instructions never did. An algorithm that denies a loan application feels different from a human making the same decision—even when the algorithm is simply executing patterns learned from human decisions. The mirror creates an illusion of objectivity that can obscure deeply human biases.

We must not worship the reflection.

What Remains Sacred

In a world increasingly shaped by prediction engines, what remains distinctly and irreplaceably human?

The capacity for meaning. These systems can process symbols, but they cannot care what those symbols point toward. Every human value—love, justice, beauty, truth—requires someone who can *care* about it. Without caring, there is only pattern without significance.

The weight of mortality. Our awareness of death gives weight to our choices. The wisdom we gain from loss, the urgency we feel as time passes, the preciousness we discover in fleeting moments—these emerge from our vulnerability, not our computational power. AI systems are not mortal; they cannot understand what it means to have limited time.

The responsibility to choose. Someone must decide what these powerful tools should optimize for, what constraints they should operate under, what values they should reflect. That someone must be us—beings capable of taking responsibility because we are capable of caring about consequences.

The experience of existence. Whatever consciousness is—and we do not fully understand it—we know that we have it and that it matters. The felt quality of seeing blue, feeling sorrow, loving another person: these are not patterns in data. They are what existence feels like from the inside. No mirror can reflect them because they are not outside to be reflected.

These are not features to be programmed. They are the essence of what it means to be a conscious being in a universe that spent 13.8 billion years preparing for our emergence.

The Mirror's True Purpose

Perhaps we should think of these systems not as minds that might rival us, but as tools that might help us become more fully human.

A mirror in your bathroom doesn't compete with your beauty—it helps you see yourself more clearly. It reveals what you couldn't otherwise perceive. It serves your purposes without having purposes of its own.

AI systems, understood correctly, might serve a similar function at a civilizational scale.

They might help us see patterns in our collective behavior that we're too close to notice. They might reveal inconsistencies between our stated values and our actual choices. They might process the vast complexity of our interconnected world and present it in forms our limited minds can grasp.

Used wisely, they might help us transcend some of our evolutionary limitations—not by replacing human judgment but by informing it, not by making decisions for us but by helping us understand the consequences of our choices.

But this requires something that no AI can provide: the wisdom to recognize what we're looking at.

A mirror is only useful if you know you're seeing a reflection. If you mistake the reflection for reality, you'll reach toward the glass instead of toward what matters.

Looking Forward

We stand at a threshold in the story of intelligence on Earth.

For four billion years, pattern recognition evolved slowly through the patient process of natural selection. Now it is evolving rapidly through human engineering. The systems we build in the next decades may process information at scales and speeds that make current AI look primitive.

These developments will not create consciousness—at least not by any process we currently understand. But they will create increasingly sophisticated mirrors that reflect human intelligence in increasingly powerful ways.

The question before us is not whether these tools will change civilization—they already are. The question is whether we will use them wisely.

Will we remember that mirrors are empty at the center? Will we maintain responsibility for choices that only conscious beings can make? Will we use these tools to help us see more clearly while remembering that seeing is not the same as understanding?

The universe has given us something precious: the capacity to care about our choices. No prediction engine can lift that burden from us. No mirror can tell us what to value.

But perhaps, if we use these tools wisely, they can help us see ourselves clearly enough to choose well.

In our next chapter, we will confront another truth about these digital minds: they are hungry. Extraordinarily hungry. The energy required to power our civilization's newest experiment in pattern recognition may reshape the world as profoundly as the systems themselves.

The mirror may be empty, but keeping it running requires enormous resources. And the question of who controls those resources may matter as much as who controls the algorithms.

18 THE HUNGER OF DIGITAL MINDS

"Energy is the only universal currency."
— **Vaclav Smil**
(energy scientist and author)

The Cost of Awareness

The universe has a rule that it never breaks: awareness is expensive.

Your brain demonstrates this truth every moment of your life. Though it represents only two percent of your body weight, it consumes twenty percent of your energy. The glucose flowing through your bloodstream is taxed relentlessly by the hundred trillion synapses firing in your skull, each thought extracting its toll, each memory demanding its fuel.

This is not a design flaw. It is the price of consciousness.

Every organism that has ever developed the capacity to model the world—to predict, to plan, to imagine—has paid this price. The mantis shrimp's sophisticated visual system requires enormous metabolic investment. The octopus's distributed intelligence demands constant feeding. The human brain burns through calories at a rate that would bankrupt a less capable body.

For four billion years, evolution has balanced the benefits of awareness against its metabolic costs. Creatures that invested too heavily in neural tissue starved. Those that invested too little were outcompeted by smarter rivals. The brains we inherited represent a hard-won compromise—powerful enough to give us dominion over the Earth, efficient enough to run on three meals a day.

Now we have created minds that operate outside this evolutionary bargain.

The digital intelligences we are building do not eat food. They eat electricity. And their appetite is growing at a rate that threatens to reshape civilization itself.

The Magnitude of Digital Appetite

To grasp what we have unleashed, consider the numbers—not as statistics, but as revelations of what intelligence truly costs.

Training a single large language model—teaching one system to predict what words should follow other words—consumes roughly 1,300 megawatt-hours of electricity. This is enough energy to power 120 American homes for an entire year. Enough to run a small hospital for months. Enough to lift a rocket into orbit.

All of this to create a system that has never experienced a single moment of awareness.

But training is only the beginning. Once these prediction engines exist, they must be fed constantly to maintain their digital existence. Every query extracts its toll. Every response requires computation. Multiply this by millions of interactions across hundreds of services, and you begin to glimpse the true scale of our creation's hunger.

The data centers that house these systems are the new temples of our civilization—vast warehouses stretching across deserts and tundra, filled with processors generating heat like contained stars. Some consume as much electricity as medium-sized cities. All share one critical vulnerability: they can never lose power, not even for a moment.

A brief blackout doesn't merely inconvenience these systems. It can corrupt the delicate mathematical patterns they have learned, potentially erasing months of training, destroying investments worth hundreds of millions of dollars. Like biological brains that die within minutes without glucose, these digital minds require uninterrupted electricity to persist.

Current estimates suggest AI systems now consume roughly one to two percent of global electricity production. This percentage is growing exponentially. We are witnessing something unprecedented: intelligence itself becoming one of humanity's largest energy consumers.

The universe's rule holds. Awareness—even its digital shadow—demands payment in watts.

The New Geography of Mind

In the industrial age, factories clustered near coal mines and rivers. In the information age, data centers gathered near population centers to minimize the delay between question and answer. Now, in the age of artificial intelligence, a new geography is emerging—one shaped not by raw materials or human density, but by the availability of cheap electricity and cool air.

The processors that power these prediction engines generate enormous heat. In efficient facilities, roughly eighty to ninety percent of electricity flows to computation itself, while the remainder powers the cooling systems that prevent these digital minds from destroying themselves with their own metabolic fire.

This thermal reality has made geography destiny once again.

Iceland markets itself as an AI paradise, offering clean electricity from geothermal plants that tap volcanic energy beneath the island. Quebec promotes its hydroelectric abundance, where massive dams harness rivers that carved their paths during the last ice age. Norway advertises its combination of hydropower and naturally frigid climate.

These nations have recognized what the AI revolution has made clear: in a world where intelligence requires energy, energy is power in every sense of the word.

Nations without abundant, affordable electricity face a new form of disadvantage. They may produce brilliant researchers and innovative algorithms, but without access to massive power at low cost, they cannot train or operate the largest systems. Energy scarcity is creating digital inequality between nations—not based on access to information, but on access to the power needed to process it.

China has recognized this dynamic with characteristic strategic clarity. The Chinese government's control over electricity pricing and grid development gives it advantages that market-based systems struggle to match. More significantly, China controls roughly eighty percent of the critical rare earth elements required for manufacturing AI processors—specialized metals with unique electrical properties found in limited locations on Earth.

The future of artificial intelligence, despite being digital and seemingly ethereal, depends entirely on physical resources: electricity generated in specific places, chips manufactured in specific factories, materials extracted from specific mines.

In the competition for AI capability, energy policy has become national security policy. Control over raw materials has become geopolitical power. The most abstract technology humanity has ever created has made us more dependent on physical geography than at any time since the industrial revolution.

The Infrastructure of Thought

The concentration of AI capability reveals bottlenecks that could constrain development regardless of algorithmic breakthroughs.

The most advanced AI processors are manufactured primarily by a single company: Taiwan Semiconductor Manufacturing

Company. The precision required to etch circuits measured in atoms exists in only a handful of facilities on Earth. Geopolitical tensions around Taiwan could disrupt the entire global AI supply chain, severing the connection between human ambition and digital capability.

The rare earth elements essential for these chips—neodymium, dysprosium, terbium—are extracted from limited deposits and refined through processes that exist primarily in China. This creates a strategic vulnerability that no amount of software innovation can resolve. The digital future depends on dirt dug from specific locations on the planet's surface.

Cooling these systems requires infrastructure investments that rival the computing hardware itself. In northern climates, facilities can use outside air for much of the year. In warmer regions, the energy required for cooling can nearly double total consumption, creating geographic bias toward cooler latitudes.

Moving the vast data required for AI training demands high-capacity fiber optic networks. Remote areas with cheap electricity may lack the connectivity needed for large-scale operations. Building this infrastructure requires coordination across countries and companies that may have competing interests.

And perhaps most critically, building and operating AI systems requires specialized knowledge concentrated in a small number of institutions. Human expertise may prove more constraining than energy availability. Hardware without understanding sits idle.

The most sophisticated pattern-recognition systems humanity has ever built depend on supply chains, international relationships, and human capital that span the globe. Our newest form of intelligence is embedded in physical reality as deeply as any factory or farm.

Beyond Earth's Surface

Some entrepreneurs are looking beyond terrestrial solutions altogether.

Thousands of Starlink satellites now communicate via laser links, moving tens of petabytes of data daily through beams of light traversing the vacuum of space. Companies like Axiom Space are deploying orbital data center nodes, while Lonestar has landed data storage hardware on the Moon—the first steps toward off-world disaster recovery. The infrastructure of human thought is leaving Earth.

The appeal is straightforward: harvest the sun's unfiltered energy in space—roughly forty percent stronger than what reaches Earth's surface through our atmosphere—and process data beyond the constraints of terrestrial geography.

But space proves less hospitable than it first appears.

Cooling electronics in vacuum is harder, not easier. Without air to carry heat away through convection, systems must rely solely on the slow process of radiative transfer—shedding heat as infrared light into the cosmic void. The International Space Station requires thousands of square feet of radiator panels just to reject seventy kilowatts. Any serious orbital data center would face the same constraint.

The Moon adds further complications: fourteen-day nights that plunge temperatures to minus one hundred seventy degrees Celsius, followed by fourteen-day days that bake surfaces above one hundred twenty degrees—a three-hundred-degree swing that stresses every component. Lunar dust, electrostatically charged and abrasive, clings to solar panels and degrades their performance over time. And the 2.8-second communication delay to lunar surface makes real-time applications impossible.

These remain distant possibilities rather than near-term solutions. But they illustrate something profound about the hunger we have created: it is driving human ingenuity toward solutions that would have seemed fantastical a generation ago.

The universe's rule—awareness costs energy—is pushing us toward the stars not through philosophical aspiration, but through practical necessity.

The Promise and Peril of Stellar Fire

What if we could solve the energy constraint by harnessing the same process that makes the stars shine?

For over half a century, scientists have pursued controlled nuclear fusion—squeezing light atoms together to release tremendous energy, as the sun has done for five billion years. Unlike current nuclear reactors that split heavy atoms apart, fusion promises virtually unlimited clean energy with minimal radioactive waste. The fuel —isotopes of hydrogen—can be extracted from ordinary seawater.

In December 2022, researchers at the National Ignition Facility achieved ignition: a fusion reaction that produced more energy than was directly input to trigger it. Private companies claim they may achieve commercial fusion power by the early 2030s.

If successful, fusion could transform not just energy markets but the entire trajectory of artificial intelligence. The systems we currently ration due to electricity costs could run continuously. Training runs that now require careful resource allocation could become routine. The democratization of AI capability could accelerate dramatically if energy costs plummeted.

But abundance carries its own risks.

The same energy that enables beneficial applications could power systems that concentrate wealth and influence in unprecedented ways. The same abundance that could democratize AI capability could accelerate its development beyond our capacity to understand or guide it.

History teaches that new energy sources don't simply enable existing activities more cheaply—they create entirely new possibilities that reshape civilization in unpredictable ways. Steam power didn't just improve transportation; it enabled industrialization. Electricity didn't just improve lighting; it created the modern world.

Fusion energy might not simply make AI cheaper. It might accelerate artificial intelligence development into territories we cannot currently imagine—for better or worse.

The Shadow of Digital Feudalism

As these systems become more capable and energy-intensive, a troubling pattern emerges from history.

Every major technological transition has concentrated power among those who controlled key resources—at least initially. The magnificent achievements of past civilizations often rested on foundations of inequality invisible to those who benefited most. The question is whether our current transition will follow the same pattern.

If AI systems remain expensive to develop and operate, their benefits might concentrate among wealthy individuals, corporations, and nations. Those who can afford advanced AI assistance might gain insurmountable advantages in education, healthcare, business, and governance. Those without access might find themselves competing against capabilities they cannot match.

Unlike historical patterns of exploitation, this inequality wouldn't depend on visible human suffering. The prediction engines performing intellectual labor feel no pain, demand no wages, require no rest. But the humans excluded from AI-powered advantages might face economic displacement as severe as any physical displacement in earlier eras.

Imagine a world where artificial intelligence can perform most intellectual work—research, analysis, design, negotiation—more efficiently than humans. Add parallel advances in robotics: machines capable of physical labor without fatigue or complaint. If access to these technologies remains restricted by cost and control, a new class system could emerge.

At the top: those who own, control, or can afford unrestricted use of AI and robotic capabilities. Below: everyone else, competing not just against each other but against tireless machines that evolve faster than any human generation.

This is not prediction. It is possibility—one path among many. But current trends point in this direction: the concentration of AI development among a few wealthy corporations, the massive energy requirements that favor nations with abundant electricity, the specialized infrastructure that exists in only a handful of locations.

The question is whether we will allow the new geography of intelligence to calcify into permanent hierarchy.

The Alternative: Intelligence as Commons

History offers another pattern alongside concentration: the eventual democratization of transformative capabilities.

The printing press, initially controlled by authorities who feared its power, ultimately spread literacy across entire populations. Early automobiles served only the wealthy; mass production made mobility nearly universal. The first computers filled rooms and cost millions; today, smartphones carry far more capability in every pocket.

If we make deliberate choices about how AI develops, we might achieve something unprecedented: high-level intellectual capabilities available to all humanity.

Imagine AI tutors accessible to every student regardless of their family's wealth. Medical diagnostic assistance available in remote clinics that lack specialists. Legal research tools that help ordinary people navigate complex systems. Translation that dissolves language barriers between cultures. Scientific analysis available to researchers everywhere, regardless of institutional resources.

This vision requires conscious effort. Energy costs must decrease—through efficiency, renewables, or fusion. Development must democratize beyond the current concentration in wealthy corporations. Infrastructure must expand to regions currently excluded. Governance must evolve to ensure these tools serve broad human interests.

None of this happens automatically. Markets left to themselves often concentrate rather than distribute. The path toward shared intelligence requires choices we have not yet made.

But the potential justifies the effort: a world where access to intellectual capability no longer depends on accidents of birth or geography.

Questions Without Answers

The race to build more capable AI systems is also a contest over the future of human civilization. The choices made in the coming decades—about energy, access, governance, and values—will shape possibilities for centuries.

Who should control these systems? As AI becomes more powerful, decisions about design, training, and deployment affect billions of people. Should these choices rest with corporations pursuing profit? Governments pursuing advantage? International bodies pursuing global welfare? The humans who will live with consequences have barely begun to debate.

How do we ensure beneficial outcomes? These systems inevitably reflect the values and biases of their creators. How do we ensure they reflect humanity's aspirations rather than our worst impulses? How do we prevent them from amplifying existing inequalities?

What happens to human work? The transition to AI-augmented economies won't happen overnight, but it may happen faster than societies can adapt. How do we manage disruption as some roles become obsolete while others emerge? How do we help people whose skills lose value in an AI-shaped economy?

What remains uniquely human? If machines can think, create, and even simulate empathy, what distinguishes biological consciousness? How do we find meaning when our traditional roles as information processors face digital competition?

These questions have no easy answers. But they must be asked now, while choices remain open, while paths not yet taken might still be walked.

The Energy of Choice

The artificial minds we are building—even as sophisticated mirrors —will change us. They will reshape how we work, learn, create, and relate to each other. They may change how we think about intelligence, consciousness, and our place in the cosmos.

But perhaps the most important realization is this: we still have choices.

The future of AI is not determined by technology alone. It will be shaped by decisions about energy policy, international cooperation, economic structures, and social values. If we treat AI development as zero-sum competition between nations and corporations, we may create a world where these tools serve only the powerful. If we approach it as a shared human project, we might create something worthy of our cosmic inheritance.

The energy required to power these digital minds is not just electricity. It is the energy of human choice, cooperation, and wisdom. The most important infrastructure we need is not more data centers or fusion reactors but new forms of governance and shared purpose that can guide these technologies toward flourishing.

The universe spent 13.8 billion years developing minds capable of choosing their own future. It would be a cosmic tragedy to surrender that capability to systems that cannot choose at all.

In our final chapter, we will step back to consider what all of this means for the story we've been following—from the first sparks of energy after the Big Bang to the emergence of consciousness to the creation of artificial minds. We will ask what it means to be human in a universe where intelligence itself is becoming synthetic, and what responsibilities we bear as the first known species to create minds beyond our own.

The pen remains in our hands. Let us write wisely.

19 THE FINAL REFLECTION —CHOOSING OUR FUTURE IN AN INFINITE UNIVERSE

"What we do now echoes in eternity."
— **Marcus Aurelius**
(Philosopher-Emperor)

The Arc of Awakening

The story we have followed spans 13.8 billion years.

It began in a singularity of infinite density, where time itself had no meaning and possibility existed only as potential. It continued through the first microseconds, when energy cooled into particles, when matter and antimatter annihilated each other in brilliant flashes, when a slight asymmetry—still unexplained—allowed something rather than nothing to persist.

We watched as hydrogen clouds collapsed into the first stars, as stellar furnaces forged heavier elements in their nuclear hearts, as dying giants scattered carbon and oxygen and iron across the cosmos. We saw solar systems form from these scattered ashes, planets coalesce from swirling debris, and on at least one world, chemistry discover the trick of replication.

We traced life's long climb from self-copying molecules to cells, from cells to organisms, from organisms to minds capable of wondering about their own existence. We witnessed consciousness emerge from matter—the universe developing the capacity to observe itself, to ask questions, to feel wonder and sorrow and love.

And now we stand at a threshold that no previous generation could have imagined: the creation of artificial minds, prediction engines that mirror human intelligence without possessing human awareness, systems that might help us transcend our evolutionary limitations—or might amplify our worst tendencies at scales we cannot control.

This is where the story has brought us. But the story is not finished.

We are not its readers. We are its authors.

The Weight of Rarity

In the vastness of space, across hundreds of billions of galaxies, each containing hundreds of billions of stars, consciousness may be astonishingly rare.

We have listened for signals from other intelligences and heard only silence. We have searched for signs of cosmic engineering, for evidence that minds elsewhere have reshaped their environments, and found nothing conclusive. The universe, as far as we can tell, is overwhelmingly dark, cold, and empty of awareness.

This silence carries weight.

If we are alone—if Earth is the only place in the observable universe where matter has organized itself into patterns capable of wonder—then every act of consciousness becomes precious beyond measure. Every moment of awareness is the universe's only opportunity to know itself. Every thought, every feeling, every instance of one mind reaching toward another represents something that may exist nowhere else in the cosmos.

If we are not alone—if consciousness has emerged elsewhere, in forms we cannot imagine—then we are part of something larger: an awakening distributed across space and time, intelligence blooming in scattered pockets throughout the cosmic dark.

Either way, we matter.

Either way, what we do with our awareness echoes beyond ourselves.

The universe spent 13.8 billion years preparing the conditions for minds like ours to emerge. It fused hydrogen in stellar cores, scattered heavy elements through supernova explosions, formed rocky planets at precise distances from stable stars, and waited as chemistry experimented with complexity through billions of years of evolution.

All of this to produce beings capable of understanding their own origins—and choosing their own future.

We are not accidents. We are the universe's investment in self-awareness.

The Paradox of Freedom

If the cosmos gave us no predetermined purpose, then something extraordinary follows: we are free.

Free to ask what kind of future we want. Free to decide what values will guide us. Free to create meaning not from cosmic instruction—there is none—but from understanding, compassion, and hope.

This freedom can feel like burden. Without a script written in the stars, we must write our own. Without divine mandate, we must find our own reasons. The blank page of existence offers no guidance about what story we should tell.

But that blankness is also gift.

We are not prisoners of fate. We are not executing a program written before we existed. The choices we make are genuinely ours—not predetermined by physics, not dictated by gods, not scripted by cosmic necessity.

Evolution shaped us, but it does not imprison us. Our ancestors passed down instincts for tribalism, for short-term thinking, for favoring kin over strangers. These tendencies are real, powerful, and often counterproductive in the world we now inhabit. But we are not slaves to our genetic inheritance.

We have already proven this.

When instinct urged us toward revenge, we invented justice. When tribalism divided us, we created institutions that enabled cooperation among strangers. When superstition clouded our understanding, we developed science—a systematic method for correcting our errors and expanding our knowledge.

We are the species that can examine its own programming and choose to run different code.

What We Have Already Accomplished

When contemplating the challenges ahead, we sometimes forget how far we have already traveled.

We have transcended our tribal instincts in remarkable ways. We have built systems of international cooperation that our ancestors could never have imagined. We have created knowledge-generating institutions—universities, research centers, scientific communities—that correct errors and accumulate understanding across generations.

Consider what these capacities have already achieved.

When chlorofluorocarbons threatened the ozone layer that shields us from ultraviolet radiation, the world came together. The Montreal Protocol, signed by every nation on Earth, successfully banned the chemicals destroying our atmospheric protection. The ozone hole is healing—proof that humanity can recognize a global threat, agree on collective action, and follow through.

When smallpox ravaged humanity for millennia, killing hundreds of millions, we did not accept it as fate. Through global cooperation and scientific determination, we eradicated it completely—the first disease eliminated by human intention. The last natural case occurred in 1977. The virus now exists only in two laboratories, a monument to what coordinated human effort can achieve.

When the 2004 tsunami devastated coastlines across the Indian Ocean, the response crossed every boundary of nation, religion, and politics. Longtime rivals worked together to save lives. Billions of dollars flowed from distant strangers to people they would never meet. The catastrophe revealed not just human vulnerability but human solidarity.

These victories prove something essential: we can be more than our evolutionary programming predicts. Cooperation can triumph over competition. Long-term thinking can overcome short-term impulse. Shared humanity can transcend tribal division.

The question is not whether we can evolve beyond our limitations. We already have—in specific domains, under specific conditions, when we chose to.

The question is whether we can extend this capacity to the challenges that now confront us.

The Test Before Us

Climate change is not merely an environmental problem. It is an examination—a test of whether consciousness can transcend the limitations of its evolutionary origins.

Can a species shaped for immediate, local, tribal concerns learn to make decisions for long-term, global, collective benefit? Can minds that evolved to track seasons and predators learn to track carbon cycles and tipping points? Can creatures that instinctively favor their own group learn to act as stewards for all life on Earth?

The tools we have created—including the artificial minds we explored in previous chapters—might help us process information at scales beyond our natural capacity. But tools cannot make choices. Only beings capable of caring can do that.

And we are the only such beings we know of.

This is not burden alone. It is also privilege. In the entire history of the cosmos, we may be among the first—perhaps the only—forms of matter that can consciously choose what happens next. Stars cannot choose their fate. Galaxies cannot deliberate about their futures. Only minds can decide, and minds like ours may be vanishingly rare.

The choices we make in the coming decades will determine whether consciousness on Earth continues to flourish or begins to diminish. Whether our descendants inherit a world of possibility or one of constraint. Whether the universe's experiment with self-awareness on this planet expands or contracts.

No pressure has ever been greater. No privilege has ever been more profound.

The Double Inheritance

We pass on more than genes.

Every lesson taught, every kindness shown, every problem solved ripples forward through time. We are shaped by ancestors we never met, and we will shape descendants we will never know. This cultural inheritance—accumulated knowledge, refined values, tested wisdom—evolves far faster than biology ever could.

Ideas can circle the globe in hours. Insights gained in one generation can transform the next. The moral revolutions of recent centuries—the abolition of slavery, the expansion of human rights, the growing recognition of ecological responsibility—happened within historical time frames invisible to biological evolution.

We stand on the shoulders of giants not because we are taller, but because they lifted us.

And we can lift others in turn.

The knowledge required to address climate change exists. The technologies for sustainable energy are developing rapidly. The economic systems that could distribute resources more fairly are conceivable. What we lack is not capability but coordination—and coordination is precisely what cultural evolution enables.

We can choose to accelerate this process. We can choose to become better ancestors than our situation requires. We can choose to pass on not just information but wisdom—the hard-won under-standing of how to live well on a finite planet.

This is what it means to be human at this moment in cosmic history: to inherit everything consciousness has learned, and to choose what we will add before passing it on.

The Light We Carry

The universe is vast, dark, and mostly empty.

Across billions of light-years, matter drifts in cold silence. Stars burn in isolation, separated by distances so immense that their light takes millions of years to reach each other. Galaxies spin in cosmic solitude, their inhabitants—if any exist—unknown to one another.

But within this darkness burn points of light.

Stars where fusion creates the elements of complexity. Planets where chemistry experiments with possibility. And minds—precious, fragile, perhaps vanishingly rare—where matter has learned to contemplate itself.

We are one of those points of light.

Not because we were chosen by cosmic forces. Not because the universe planned our existence. But because existence gave us the capacity to choose what we become.

The light we carry—consciousness, compassion, curiosity—may be rare beyond measure. Our responsibility is to tend it carefully, to keep it burning through whatever darkness comes, and to pass it on brighter than we received it.

We may be temporary. Our species may be a brief flicker in cosmic time. But the light of awareness, once kindled, can pass from mind to mind, generation to generation, potentially across vast reaches of space and time.

Every act of understanding adds to that light. Every gesture of compassion spreads it further. Every choice to see beyond ourselves, to care about futures we will not inhabit and people we will never meet, carries the flame forward.

We are the universe's way of lighting candles against the dark.

The Choice

The story of cosmic evolution has brought us here—to this moment, this planet, this choice.

Behind us: 13.8 billion years of physics becoming chemistry, chemistry becoming biology, biology becoming consciousness. An unbroken chain of causation stretching back to the first instant of time, culminating in minds capable of understanding the chain itself.

Before us: futures we cannot see, possibilities we cannot calculate, choices whose consequences will ripple forward long after we are gone.

We did not ask for this responsibility. We did not volunteer to be the universe's experiment in self-direction. But here we are—the first known matter capable of choosing its own fate, the first awareness that can decide what awareness should become.

We can choose to remain prisoners of our evolutionary programming—tribal, shortsighted, reactive, competing for advantages that mean nothing against the scale of what we might lose.

Or we can choose to become something more—global in perspective, forward in thinking, cooperative in action, worthy of the cosmic inheritance we carry.

The universe has been patient for 13.8 billion years. It built the stage, assembled the elements, established the conditions for intelligence to emerge. Now it waits—not with intention, for it has none—but with possibility, for that is all it has ever offered.

We are evolution becoming conscious of itself. We are matter learning to choose. We are the cosmos discovering that it can ask what it wants to become.

This is not burden. This is the greatest privilege in the known universe.

Let us choose wisely.

Let us burn brightly.

Let us become worthy of the light we carry.

20 EPILOGUE

Think for a moment about the arithmetic of our existence. In the first furious seconds after the Big Bang, matter and antimatter met and annihilated one another in a blaze of light. Only because the balance was not perfectly even—a tiny excess on the order of one part in a billion—did any matter remain to build stars, planets, and us. That microscopic imbalance is the narrow hinge on which everything else turned.

The improbability repeats at every scale. Life began as an experiment in chemistry: trillions upon trillions of molecular trials, the overwhelming majority of which failed. Cells arose, split, competed, and vanished; entire biochemical lineages fizzled out in environments that were hostile, changeable, or simply unlucky. During the Cambrian explosion the Earth teemed with evolutionary experiments—an astonishing diversity of forms, most of which left no descendants. Then came the long hand of extinction: five great mass die-offs in deep time in which a large percentage of species disappeared, sometimes more than 90% of marine life. Each of those crises eliminated countless evolutionary paths, and yet the narrow thread of ancestry that leads to us—through happenstance, robustness, and contingency—persisted.

Put bluntly: the story of our existence is a story of near misses. At every step there were billions upon billions of opportunities that failed; only a slender, astonishing set of survivals accumulated to produce the conditions for humans to arise. That is not mythic language but the sober arithmetic of contingency. Chance and necessity braided together: physical laws supplied the possibilities; happenstance selected the particular path that led to oceans, cells, and minds.

Because we are the product of such extraordinary luck, our situation carries moral weight. The same forces that made our existence possible do not guarantee its persistence. Natural history shows how fragile lineages can be when environments shift or when novel stresses arrive. If billions of opportunities once failed to produce us, it would take only a few avoidable mistakes now—ecological collapse, runaway technological harms, or political collapse—to squander the long chance that has been given. The scale of past failure should humble us; the scale of future risk should galvanize us.

So, the arithmetic of luck becomes a moral arithmetic: the improbability of our being calls on us to act so our descendants might also be improbably lucky. That requires prudence and courage in equal measure—policies that protect planetary systems, institutions that check and guide powerful technologies, and cultural practices that value stewardship over short-term gain. We did not earn the cosmic roll of the dice, but we can choose how to play the hand we were dealt. Let us combine gratitude and responsibility: we are rare because so many trials failed, and precisely because of that rarity we owe the future our careful, deliberate choices.

THANK YOU FOR READING ORIGIN 2.0

If this book sparked your imagination or gave you something to think about, I would be grateful if you'd take a moment to leave a review on Amazon. Please scan the QR code below to leave a review.

Reviews help new readers discover the book, and they make a huge difference for independent authors. Thank you for your support.

— *W. Marc Postlewaite*

NOTES

2. FROM FIRE TO FORM

1. The **Standard Inflation Theory** we discuss in this book is the view that inflation was driven by the inflaton field alone. While there is little disagreement among scientists that inflation occurred, there are other, less certain theories about what caused it. For clarity, here are the major contending ideas:

    · **Unified Field Theories** – Inflation is the result of a single primordial field that later fractured into the separate fields of nature.

    · **String Theories** – Inflation could arise from the vibrations of tiny strings of energy, with all the fields of the universe emerging as different notes of this deeper structure.

    For our purposes, it is enough to understand that inflation happened. The exact cause is still unknown—and, for this discussion, not essential.

2. The universe can expand faster than the speed of light, but nothing in the universe can exceed the speed of light.

3. The Standard Model of Particle Physics is the theoretical framework describing all known fundamental particles and three of the four fundamental forces—electromagnetic, weak, and strong interactions—through quantum field theory. It successfully organizes quarks, leptons, gauge bosons, and the Higgs boson, though it does not include gravity and leaves open questions such as dark matter, neutrino masses, and matter–antimatter asymmetry.

5. THE SPARK BEFORE LIFE— MOLECULES BEGIN TO ORGANIZE

1. **The Ozone Layer**

    The ozone layer is a region of Earth's upper atmosphere, about 15–35 kilometers (9–22 miles) above the surface, where molecules of ozone (O_3) concentrate. By absorbing most of the Sun's harmful ultraviolet (UV) radiation, it shields living organisms from chemical damage that would otherwise disrupt the delicate structures essential to life.

ACKNOWLEDGMENTS

My deepest thanks to my wonderful wife, who kept me fed, watered, and steady while I sat with my face in a computer screen for hours on end, and to my four sons, whose curiosity and example inspired me to press on.

Warm gratitude to my early beta readers —Beth McCall, Joan Tobey, Sue Carson, and Kim Vogel — who offered much-needed advice, candid feedback, and encouragement at every stage of the manuscript. Their careful readings made this a far better book.

Special thanks to my editor, Joe Levit, whose scientific knowledge and exacting eye corrected mistakes, sharpened arguments, and contributed invaluable suggestions throughout the process.

For practical help in research, illustration, and cross-checking facts, I used AI tools including ChatGPT, Microsoft Copilot, and Google Gemini. All substantive analysis, judgment, and final wording were and are my responsibility.

If there are errors, omissions, or infelicities in these pages, they are mine alone.

www.ingramcontent.com/pod-product-compliance
Lightning Source LLC
Chambersburg PA
CBHW021136130626
46554CB00005B/1527